BLACK PATHS AND GREEN CATHEDRALS

BLACK PATHS AND GREEN CATHEDRALS
A Guide to Ecological Paganism

Sian Sibley

AEON

First published by Black Lodge Publishing 2023
Second edition by Aeon Books 2026

Copyright © 2026 by Sian Sibley

The right of Sian Sibley to be identified as the author of this work has been asserted in accordance with §§77 and 78 of the Copyright and Patents Act 1988.

All rights reserved. No part of this publication can be reproduced, stored in retrieval system, or transmitted, in any form or by any means, electronic, mechanical, photocopying, recording or otherwise without the permission of the author and publisher.

British Library Cataloguing in Publication Data

A C.I.P. for this book is available from the British Library

ISBN-13: 978-1-80152-226-7

Typeset by Medlar Publishing Solutions Pvt Ltd, India
All illustrations produced by Sian Sibley

www.aeonbooks.co.uk

*I dedicate this book to my
Brothers and Sisters in the Green.*

*The unseen but ubiquitous,
The invisible providers of all we have,
The unacknowledged majority Standing in plain sight,
The Green and Glorious Mighty Trees.*

CONTENTS

PREFACE TO THE SECOND EDITION   xi

INTRODUCTION   xiii

    PART ONE: WHEN THE EARTH BECAME OTHER

CHAPTER ONE
Tracing the faultlines: The roots of ruin   3

CHAPTER TWO
Paganism: The great green hope   18

CHAPTER THREE
An Oxymoron: 'Ecologically Unsound Paganism'?   29

    PART TWO: BLACK PATHS AND GREEN CATHEDRALS

CHAPTER FOUR
Working with the land spirits   61

PART THREE: GREEN CATHEDRALS: THE MAGICAL TREES

CHAPTER FIVE
Working with Charubel's plant sigils — 113

CHAPTER SIX
The Oak — 117

CHAPTER SEVEN
The Yew — 130

CHAPTER EIGHT
The Ash — 144

CHAPTER NINE
The Rowan — 154

CHAPTER TEN
The Alder — 170

CHAPTER ELEVEN
The Birch — 184

CHAPTER TWELVE
The Holly — 194

CHAPTER THIRTEEN
The Hawthorn — 202

CHAPTER FOURTEEN
The Blackthorn — 210

CHAPTER FIFTEEN
The Willow — 218

CHAPTER SIXTEEN
The Hazel — 228

CHAPTER SEVENTEEN
The Ivy — 236

CHAPTER EIGHTEEN
The Apple — 250

WALKING IN THE GREEN — 262

BIBLIOGRAPHY — 267

PREFACE TO THE SECOND EDITION

When I first wrote *Black Paths and Green Cathedrals*, it came from a place deep within me—a place where love for the land, anger at its destruction, and reverence for the spirits of the wild all burned together. It was, and still is, the book of my heart. If I write a hundred more books in my lifetime, I suspect this will remain the most important one. For me, this work is not just about magic, nor even just about Paganism; it is about survival—not only for ourselves but for all that lives and breathes alongside us.

So why a second edition? Because I could not leave it as it was. The land has more to say, and so do I. Since the first edition, I have written my thesis on the ecological aspects of Paganism. This work took me deep into the ways modern Pagans are engaging (or failing to engage) with the environmental crisis. That research has given me both hope and concern in equal measure.

The hope lies in the fact that many in our community are quietly and determinedly living as if the Earth truly matters. They are planting for pollinators, collaborating with local communities, reducing waste in rituals, fighting to protect sacred places, and reminding us that reverence is an action, not just a feeling. But I am also concerned, because there are still too many of us who speak of loving the Earth while continuing to consume, pollute, and treat her as a resource rather than a living being.

Paganism, at its core, offers us something rare and precious: a chance to step out of the anthropocentric disaster we are currently living through. It provides us with a worldview in which humans are not the centre, not the rulers, but kin among kin in a vast, sacred community of life. If we are brave enough to live that truth—really live it—Paganism could be one of the most important movements of this ecological age.

But it will not happen by accident. It will not happen if we continue to talk about "honouring nature" while importing endangered herbs, throwing ritual waste into rivers, or buying spirituality in plastic packets. It will happen only if we are willing to change—to take a hard look at our practices, strip away what harms, and rebuild our Craft in right relationship with the Earth that sustains us.

That is why this second edition exists. I have added new material, not only to refine ideas but to push them further—to challenge you, as I continue to challenge myself, to look again at what it means to walk the witch's path in a time of ecological crisis. This is not an abstract philosophy; this is a living, urgent work. The land does not have time for us to be comfortable.

I believe Paganism has the power to help us reimagine our place in the world, not as masters or consumers, but as participants in a great web of life. It is a chance to remember that the Earth is not a backdrop to our human story; she *is* the story, and we are only one of her many voices. If we can live as if that is true, if we can put it into every ritual, every choice, every breath, then perhaps we can begin to heal not just ourselves but the world we belong to.

This book is my offering to that work. It is my prayer, my challenge, and my love letter to the green-blooded ones who have been my greatest teachers. I hope you will walk these paths with me—for the trees, for the rivers, for the stones, for all our wild kin, and for the future that still might be.

Walk in the Green.

Sian Sibley
2025

INTRODUCTION

We live on a planet that is, quite simply, a living miracle. As far as we know, this is the only place in the universe where Life pours itself out in such wild, diverse, and astonishing forms. It creeps and climbs, swims and soars, roots and blossoms in every crack and corner of the Earth. Life isn't rare here, it's relentless.

But, much to our shock, Humans are not the only people who live here. Not by a long shot. We're just one species among countless others, part of a vast, ongoing dance of expression that this planet hosts so generously. When you start to really see that, you realise Earth isn't just a backdrop for life, it *is* Life. She's not a rock we live on, but a being we live *with*, and that's what makes her so breathtaking.

The more you learn about our place in this world, the clearer it becomes that we are intricately entwined in a vast and complex web, and within it, our task isn't to dominate, but to live; to experience what it means to be truly *part* of Nature. That, I believe, is reason enough for our being. We are meant to be in symbiosis, not supremacy.

However, human life, for all its beauty, has its drawbacks. On one hand, we can breathe life into the world through care, kinship, art, and imagination. On the other, we can become poison, a force capable of unravelling the very Earth we were born from.

We're gifted with a fire, ferocious and generative, that has shaped language, music, story, and science. It's astounding really, and very beautiful. And yet, a shadow walks beside it. There's a lingering, dangerous belief, held tightly in some quarters, that we sit at the pinnacle of creation. That we've been handed dominion by divine right. And from that springs a careless arrogance: the idea that human wants must always come first, no matter the cost to the rest of the living world.

The idea that humans stand at the centre of life on Earth, that 'God delusion', still haunts much of our thinking. It paints us as the pinnacle, the chosen species, the ones who matter most. But I would argue otherwise. The real life-givers, the true centre of this planet's breath and being, are the trees and their green kin.

It is not us but the green ones who hold the power to create. Trees, plants, algae, and even the humble photosynthetic bacteria are the weavers of matter itself. They spin the fabric of our physical world from nothing more than sunlight, water, and an invisible gas. That is alchemy on a scale far beyond human hands.

Of course, the science behind this is more intricate than the poetic truth alone. Photosynthesis is no simple magic trick; it is a marvel of biochemistry. In the green heart of plant cells, the little organelles called chloroplasts contain chlorophyll. This biodynamic molecule interacts with light energy and initiates the light reactions, producing two incredible biological molecules: adenosine triphosphate (ATP) and nicotinamide adenine dinucleotide phosphate (NADPH). These are the tools the plant uses to carry out the next stage, known as the Calvin Cycle, where carbon dioxide from the air is fixed, converted into solid matter, and transformed into glucose, the basic sugar of life.

It's worth pausing to grasp the scale of this. Photosynthesis takes place in almost every green part of a plant, from the leaves to the stems and even unripe fruits. In the leaves, especially, the chloroplasts gather in such density—around half a million per square millimetre—that they form a living solar panel. Each tiny cell, each flicker of green, quietly performs this ancient act of creation.

The basic equation is deceptively simple:

$$6CO_2 + 6H_2O + \text{light energy} \rightarrow C_6H_{12}O_6 + 6O_2$$

But from that, our entire world unfolds. Every breath we take, every apple, every whale, every drop of blood in our veins owes its existence to this quiet, constant miracle.

Green plants, trees, and algae are the only ones that create matter from light; they actually complete Einstein's theory, where $E = MC^2$. They are the weavers of form, the architects of substance. All other beings, ourselves included, are simply consumers. We eat the green ones, or we eat other consumers who have eaten the plants. And so, we move in circles: an alchemical *circulatoria*, where life feeds on life, and all returns to the green.

Trees, in particular, are not just important; they are essential. Their role in the ecosystem is profound, multifaceted, and far too often overlooked. Yes, their beauty is staggering, but at present, they are only given value commensurate with the price tag assigned to them in our economies. However, to see a tree only as scenery or timber is to miss the heart of the matter.

Trees shape the very breath of the world. They cleanse the air of pollutants and fine particles, offering a kind of silent medicine to both human lungs and the lungs of the Earth. As carbon sinks, they hold within their trunks and roots the carbon that would otherwise heat our atmosphere, softening the blow of climate change with every growing ring.

But their gifts don't end with air. Trees hold the land in place. Their roots grip the soil, keeping it from washing away in storms; their fallen leaves nourish the earth, building fertile ground for new life. They offer food and shelter to a multitude of beetles, owls, foxes, fungi, and us. Their branches cradle nests; their bark hides countless micro-worlds. They are the homes and the hosts.

Even the water listens to trees. Through their roots, they absorb rain and rising groundwater, helping to slow down floods. Through transpiration, they allow water vapour into the sky, forming clouds, calling down rain, and regulating the great cycle of water that all life depends upon.

One of the most vital roles trees play, and one that often goes unappreciated, is their quiet yet powerful work in regulating the climate. Forests shape both local weather and global patterns; they soften temperature extremes, guide rainfall, and stabilise regional conditions. In short, the green people hold the balance. Without them, the web begins to unravel.

Trees are not background; they are participants, companions, and teachers. They are the alchemists of air and water, the quiet architects of balance.

So, if there is a force at the centre of life on Earth, it is not us. It is the green ones who create with sunlight and feed the world.

Let's be honest: the ecological crisis we now face has not come from the trees, or the waters, or the air, it's come from us. In the last two centuries alone, we've orchestrated mass deforestation, driven species to extinction, shifted the very climate, and polluted air, soil, and water with chemicals and plastic. And somehow, each day seems to bring a fresh absurdity, new ways of poisoning the world that gave birth to us.

So the question stands: what is humanity's grand solution to this slow suicide?

Well, sadly, the answer appears to be to leave the planet.

We seem to want to do that in two separate ways. The first way is to die, to leave the planet and go to Heaven.

The long shadow of religious thinking teaches us that this world is only temporary, a test, a trial, something to endure on our way to a 'better place'. For centuries, we've been told that after death, we'll be swept up to a heavenly realm. In this glittering paradise, we can sit at the right hand of some cosmic patriarch and live forever in comfort. Earth, in this story, serves as a disposable backdrop to the real show, which unfolds after you die.

And if that's the story you believe, then why not strip the world bare? Why not take and take until there's nothing left? Why care for a planet that's little more than a pitstop before the real thing starts?

This is the slow rot, a bone-deep sickness seeping through the cracks of our world. It is a Cult of Death, a mindset that worships endings and feeds on ruin. It stockpiles weapons for profit, devours everything in its path, and leaves forests razed, oceans poisoned, and futures gutted in its wake. You can see it everywhere—in our love affair with war, our giddy hunger for destruction, our obsession with violence neatly packaged as entertainment.

It's there in how we gorge on meat without thought, shoot animals for fun and dare to call it 'sport'. This cult has taught us to forget how to truly live; instead, we exist in a constant state of paranoia, fear of war, fear of disease, fear of never having 'enough'. And here lies the trap: greed and gluttony are not just permitted, they are *encouraged*, driven by a profiteering media that feeds our fears because fear sells.

And yet, no one ever tells us what 'enough' even means. We are kept hungry on purpose, starved of contentment, because a person who knows they have enough is a person who cannot be controlled.

Then there's the second fantasy, the one shaped not by dogma but by science fiction. The idea that somewhere, out there in the stars, lies a better world. A cleaner slate. A planet we haven't ruined yet. And so we pour fortunes into scanning the sky, searching for another Eden, while ignoring the one we're already living in, right outside the window.

But here's the thing: if we are to have any future at all, if we're to survive as a species *and* truly thrive, we must learn to live differently. We must remember how to walk in step with the Earth and all her beings, to live in reciprocity with our green kin. That is the work. Not escape. Not domination. Not delusion. It is the forming of relationships.

This book isn't a magic wand. It won't change the world just by being read. But if you're willing to do something more—if you're ready to take what's offered here and carry it into your life, your practices, your choices—then yes, it can have power. True magic isn't passive. It's participatory. It takes root in action.

What follows are practical teachings and magical pathways—ways to reconnect with the Earth, to work with the spirit of the land, the plants, the stones, the animals, and with Nature herself. She who is not just Earth but *Life*, a living, breathing, radiant being whose generosity humbles even the stars.

Now, I won't lie to you; some bits of this book will challenge you. Some bits might even annoy you, maybe even piss you off. And that's a good thing. Because this isn't just a feel-good handbook for lighting candles and whispering to your Gods. It's a call to look at your practice and your life, square in the eye. It's about asking hard questions, digging into uncomfortable truths, and being willing to shift, to change, to grow. Real magic isn't always sweet. Sometimes, it bites. Sometimes it breaks. But if you let it, it also heals, rebuilds, and re-roots you in something real.

So take what you find here and walk with it, into the wild, into the garden, into your own soul. Make the changes, use the magic, tend the land. And remember: you're not doing this alone. You walk with the Green.

Let's begin.

Do We Walk with Dead Hearts?

Do we walk with dead hearts,
Through a world straining to speak to us?
Consuming like zombies, mad with want,
Jealous of those who have what we are told we need.
We tell our children, You can be anything,
Never understanding that simply being is enough.
We walk through the world with closed eyes,
Seeing only human life.
We feel the rain and run for the umbrella.
We feel the cold and turn the heating up.
We stand in wild green forests and ignore the Elder people.
We look at the ocean and see fish fingers,
and a place to dump our filth.
We love the invisible God who watches us,
Who makes us feel important, and ashamed,
Who washes us clean after we dirty ourselves in His name.
But we do not love the spirits of the land
the trees, the rocks, the fire and the others
For they are taught to us as servants,
Things without worth of their own.
God says so.
Science, the new God in its white coat, says so too.
The great mechanistic judge of reality
Ignoring any experience it cannot measure.
A new inquisition, wielded to limit and remove.
We cannot feel the joy of pure being,
For we are no longer of the world.
We are users of it.
We do not speak to the other people
the green people, the stone people, the river people
Because we have convinced ourselves they are not real.
We drive our dinosaur-fuelled cars,
boxed away from the world,
Afraid it might ask something of us,
Ask us to share.

We feed our drone-eyed children.
plastic food and poisons,
And we profit from their sickness.
More, we are told.
We must have more.
If there is no growth, we are doomed.
But in constant growth, we doom ourselves.
We never say that enough is all you need.
That sharing is joy.
That community is all beings,
Not just those we see and use.
How do we begin?
By acknowledging their lives,
Their right to exist beyond our needs.
How do we speak to them?
We say it is possible. We listen.
Seeing is not believing.
Believing is seeing.
Believe—and be free.
See—and be awed.

PART ONE

WHEN THE EARTH BECAME OTHER

CHAPTER ONE

Tracing the faultlines: The roots of ruin

I've no doubt that you feel it too, that sense that, as a species, we've become profoundly disconnected from Nature. You can see it everywhere. With the rise of mental health disorders. In children who no longer know where their food comes from or what season we're in. In the way we treat water as if it were just a utility rather than the sacred lifeblood of the Earth. Our roots have been severed, or perhaps more accurately, the roots are still there, but we've forgotten how to feel them.

There are many reasons for this disconnection, but one of the most insidious is the belief that humans stand apart from the rest of life, even above it. This isn't just a matter of personal ego; it's a cultural doctrine. To such an extent that scientists have named our current era the Anthropocene, an age characterised by the profound impact of human activity on the planet. We've carved our mark into the very geology of the Earth.

At the heart of this is *Anthropocentrism*, a worldview that places humans at the centre of the cosmos, assigning us a superior, elevated status over all other life. This human-centric lens prioritises our needs, desires, and comforts above all else. It pushes everything else, beast, tree, river, fungus, to the edges of consideration, as if they are secondary, expendable, or merely 'resources'.

This way of thinking has become so deeply ingrained in our societies that it has shaped everything from ethics to politics. It's embedded in how we govern, how we plan, how we justify destruction in the name of progress. You don't have to look far. Take, for instance, the recent scrapping of the UK's commitment to the Green Deal. Or the stubborn refusal to part with fossil fuels, to give up our petrol-driven comforts. These are not just policy decisions; they are symptoms of a deeper belief system: that humans come first, no matter the cost.

It's an ideology that thrives on short-term thinking and keeps us stuck. It trades planetary health for popularity and long-term resilience for short-term power. And worst of all, it stops us from coming together with any real sense of unity or responsibility when the Earth most needs our loyalty.

If we are to move forward, we must challenge this thinking at its roots. Because until we shift from a mindset of supremacy to one of relationship, no policy, no protest, no magic will be enough.

The damage we're doing to the land, the sea, and the very air we breathe is plain as day. You'd have to be asleep, or in politics, not to see it. And yet, here we are, drifting along in a shared illusion, led by people who cling to denial like a lifeboat. Change is inconvenient, especially when profit's involved.

This human-centred, 'we're-the-main-character' mindset isn't new, either. You can see the roots of it back in Classical Greece. Anyone who wasn't Greek was 'Barbaroi', even if they were Egyptians with libraries older than Athens. The arrogance ran deep. From this grew the so-called Great Chain of Being. This theosophical model began to take shape as early as the 8th century BC in the Homeric epics, such as the *Iliad*, and was refined through the Classical and Hellenistic periods. By the time we reach the Byzantine era, it had become central to Christian thought.

In this cosmic hierarchy, God sits neatly at the top. Everyone else is stacked below, and their value is assigned by proximity to the divine. The closer you are to God, the more important you are. Simple. Convenient and totally anthropocentric.

This model has rightly been criticised by Christian scholar Lynn White Jr. who ascribes anthropocentrism to a quotation from the Bible, whereby God gave Adam control and dominion over the animal and plant worlds. White notes that:

> By gradual stages, a loving and all-powerful God' had created light and darkness, the heavenly bodies, the Earth and all its plants, animals, birds,

and fishes. Finally, God had created Adam and, as an afterthought, Eve to keep man from being lonely. Adam named all the animals, thus establishing his dominance over them. God planned all of this explicitly for man's benefit and rule: no item in the physical creation had any purpose save to serve man's purposes. And, although man's body is made of Clay, he is not simply Part of Nature: he is made in God's image. Especially in its Western form, Christianity is the most anthropocentric religion the world has seen.[1]

As a scholar not afraid to court controversy, Lynn White would go on to offend many of his academic contemporaries, especially Christians who had read his paper at the time. The latter were highly offended by what he had to say. White argued that biblical doctrine in the Book of Genesis had allowed the wholesale exploitation of environmental systems on a global and historical scale. The verse that promoted this problem was in Genesis 1:28:

> *And God said to them: "Be fruitful and multiply and fill the earth and subdue it, and have dominion over the fish of the sea and over the birds of the heavens and over every living thing that moves on the earth."*

Genesis 1:28 is often referred to as a cultural mandate, which is said to give Christians divine authority. It is in this that the original call to stewardship is given.[2] By this command, man can therefore use and abuse the natural world as he sees fit.

However, in my opinion, this is a massive misunderstanding of what the word 'stewardship' actually means. In the Christian definition, it means the right to use and take from the environment as humans see fit. However, my definition of Stewardship is one of value, respect, and reciprocity.

Val Plumwood, a fierce voice of ecofeminism, exposed how the split between humanity and nature wasn't just philosophical, it was political. Rooted in patriarchal religion and Western thought, this dualism lifted 'man' and 'mind' above 'woman', 'nature', and 'emotion'. Reason was

---

[1] Lynn White, 'The Historical Roots of Our Ecologic Crisis', *Science*, 155 (1967): 1203–07.
[2] Hugh Whelchel, 'Three Key Bible Passages About Stewardship', *Work and Economics Institute of Faith* (2023).

made male. Feeling was associated with femininity, and reason and intellect were consistently ranked higher.[3]

This hierarchy has done centuries of damage. It sanctified oppression, justified conquest, and drowned out any voice not cloaked in male, white, intellectual authority. Women, Indigenous people, and the Earth herself were cast as the irrational, the emotional, the lesser, and punished for it. If your knowing comes from the heart, it's hysteria. If it comes from the land, it's superstition. And if it challenges power, it's heresy.

This is not just history. It's the architecture of the crisis we live in now.

A further profound fracture in our relationship with the natural world emerged with the rise of the Mechanistic Universe Theory. This worldview was simultaneously adopted by both the scientific and religious establishments. By the early eighteenth century, rational theology had become the dominant lens. God was no longer encountered through mystery, revelation, or direct communion with the sacred, but through human reason, logic, and controlled observation. The natural world was dissected, categorised, and boxed into theory.

This shift allowed experiment and measurement to become the supreme authority in science. Theologians of the time also welcomed it. A mechanistic cosmos seemed to echo their vision of a rational God, one who stood outside of creation like a divine clockmaker, constructing a lifeless machine to run according to fixed laws. Nature, in this model, was not a living presence but a set of parts; not kin, but object.[4]

In this vision, God became the distant engineer, and the world His automaton. The sacred was no longer immanent in river or stone, leaf or beast. Spirit was divorced from matter. The Earth was no longer a being to revere, but a mechanism to exploit.

This ideology carved a gaping wound in the Western psyche. It severed us from the breathing world, replacing relationship with control, and wonder with dissection. We are still living in the ruins of that disconnection. We see it in collapsing ecosystems, in sterile industrial landscapes, in the hollow eyes of a culture that has forgotten how to belong

---

[3] Val Plumwood, 'Gender, Eco-Feminism and the Environment', in *Controversies in Environmental Sociology*, ed. Rob White (New York: Cambridge University Press, 2004), pp. 43–61.
[4] William B. Ashworth, 'Christianity and the Mechanistic Universe', in *When Science and Christianity Meet*, ed. David Lindberg and Ronald Numbers (Chicago: University of Chicago Press, 2008), pp. 61–84.

to the Earth. The mechanistic myth may have promised understanding, but what it delivered was estrangement.

The idea of owning the land

The disconnect between the idea of *'being* in the land' and *'owning* the land' began millennia ago but was exacerbated by the Enclosures Act (1773). This act was initiated to allow 'more use of land for agriculture'; however, what it did was to allow the rich and influential to claim the land as their own.

In medieval times, farming was based on large fields, known as open fields, in which individual yeomen or tenant farmers cultivated scattered strips of land. Then as early as the 12th century, agricultural land began to be enclosed.

This meant that holdings were consolidated into individually owned or rented fields.

Initially, the enclosure of land was often a fairly informal affair; local agreements were made between landowners and farmers. But by the 17th century, the tide began to turn, and people started seeking formal permission through Acts of Parliament. Usually, it was the landowners pushing for it, hoping to squeeze more rent out of their estates, or ambitious tenant farmers seeking to increase the productivity of their farms. From the 1750s onwards, though, Parliamentary enclosure became the norm.

Between 1604 and 1914, over 5,200 Enclosure Acts were passed, affecting just over a fifth of England's land—roughly 6.8 million acres. That's a vast change, and while it undeniably boosted agricultural productivity, especially in the later 18th century, it came at a cost.

Communal open fields, which had shaped village life for generations, were parcelled up and fenced off. The old shared landscapes vanished, and with them, a whole way of living and relating to land. What was once open and collectively worked became fragmented and privately held, reshaping both the countryside and the communities who lived within it.

At the very bottom of rural society were the poor and the indentured, especially agricultural labourers, who were often forced to leave the land for good. Many had no choice but to move into the growing towns, searching for factory work or any way to survive. These were not strangers to the land; they were its original keepers, people whose families had worked and lived with the soil for generations. Their bond with the land was deep and enduring, woven into their identity, their stories, and their daily lives.

Once they were removed, that connection was severed. The land, once seen as kin, as home, as sacred, came to be viewed simply as property, something to be bought, sold, and profited from. A shift took place in how people related to place. Ownership replaced relationship. The land no longer held many in its embrace, but instead became the possession of a few.

That wound has not healed. We still live with the consequences of this separation. The disconnection between the wealthy and the poor, as well as between people and the natural world, can be traced back to these moments. What was once shared was enclosed. Not just with fences and legal papers, but in the very way people thought about land and belonging. And with that came a disconnection, not only from the Earth beneath our feet, but from one another.

Indigenous concepts of land ownership, worldwide, differ significantly from those in the West. Often, they prioritise communal or collective ownership and stewardship. The land is viewed as belonging to the entire tribe, rather than to individual members of the community. Decisions regarding land use, management, and conservation are made collectively, taking into account the well-being of the community and future generations. They are part of an ongoing relationship with the land that extends beyond their own lifetime. They must preserve and pass on the land, resources, and traditional knowledge to future generations. The land, in this respect, is therefore intricately tied to the indigenous populations' identity, culture, and way of life. The land provides sustenance, habitat, and the foundation for their traditional practices, ceremonies, and livelihoods.

Before the Enclosure Acts, people in the West could still live in rhythm with the seasons. The land was not something apart from us—it was part of us. Plants, animals, rivers, and trees were treated as integral members of the wider community, not simply as resources to be exploited. However, once ownership of the land was transferred to wealthy landlords and placed under government control, something fundamental was lost. We became strangers to the very soil that had once cradled us, and that separation has cost us more than we often care to admit.

Of course, humans have always worked with the land. We have always needed to grow food, build homes, and raise families. However, there was once a kind of balance, a living relationship, where the needs of other beings were also considered. It was not perfect, but it was closer to a mutual exchange than what we see now. Today, it is as though we have stopped seeing the land altogether. It has become a backdrop, a stage for real estate, leisure, or profit. Not a presence. Not a home, and most importantly, not considered alive in a sentient way.

So we must ask, what else changed? Yes, we have enclosure and industrialisation. Yes, the rise of monotheistic religion, with its heaven-above and dominion-below worldview, certainly played a part. But that cannot be the whole story. What else severed the ties between people and place? When did we stop listening to the land? When did we decide that control was more important than kinship?

These are questions worth asking, not to lay blame, but to begin the long journey back to remembering. Back to relationship. Back to reverence.

Is money the root of all evil?

The old saying claims that money is the root of all evil, but that simply isn't true. Money, in itself, is nothing; just an agreed-upon symbol. This collective hallucination only works because we all choose to believe in it. A banknote has no power without that shared belief. Gold does not whisper greed into our ears. Coins do not tell us to hoard. It is *people* who twist money into something monstrous.

The real root of evil lies in greed, in that grasping hunger always to have more than your neighbour, in the belief that your worth can be measured in possessions. This sickness exploded in the 1980s when we were sold the poisonous mantra that 'greed is good', and 'community is weakness'. Many of us bought into it. It is not money urging you to keep up with the Kardashians or to measure your happiness in shopping bags; we do that to ourselves. We choose to play that game, even when it robs us of joy.

The real antidote is shockingly simple, though terrifying to a culture addicted to consumption: define your own 'enough'. Not society's version, not what the adverts tell you, not the 'success' sold to you in magazines, but your personal, soul-deep *enough*. Because once you know what enough is, you are no longer easily controlled. You are no longer feeding the machine. Contentment, not wealth, is the greatest act of rebellion against this system.

Capitalism is an economic and social system driven by the pursuit of profit. At its core lies private ownership of land, resources, factories, and wealth. Individuals and companies control these resources, deciding how they are acquired, used, and traded. The marketplace becomes a battlefield shaped by competition and ambition.

On paper, capitalism sounds neutral, even fair. People and businesses have the legal right to own land, buildings, and tools of production, and to buy and sell as they see fit. However, in practice, this obsession with ownership and profit has given rise to deep-seated inequality. When economic gain takes precedence, people, communities, and the living world are pushed aside—or sacrificed entirely.

And here we reach the heart of the matter: **how can anyone claim to own land or water?** These are not lifeless objects. They are living, breathing parts of the Earth, ancient and sacred. And yet capitalism treats them as commodities, things to buy, sell, and wring dry for every last drop of profit. A company can 'own' a river. A government can 'own' a forest. But where is the care? True ownership should mean stewardship, guardianship, and love. Instead, land is exploited, rivers are poisoned, and forests are stripped bare.

We hear words like 'acquire', 'use', and 'dispose of' as if they are harmless, neutral words. They are not. They are words of domination. They show how far we have drifted from right relationship. The land is not a commodity. It is not ours to ruin.

This mindset places humanity at the centre of everything, assuming that our wants are always more important than the needs of animals, plants, water, and soil. So we pour fertilisers onto crops to force unnatural yields, we strip fields bare with monoculture farming, and soil that once teemed with life turns to dead dust. Food may still grow—for now—but at what cost? The wild is pushed out, diversity lost, and the Earth is left gasping so that profit margins can rise.

This is not ownership. This is not care. This is disconnection at its most dangerous.

In capitalism the pursuit of profit is the fundamental driving force. Individuals and businesses aim to generate income and accumulate wealth by producing goods or providing services that consumers demand. Thus, pursuing profit is one of the most significant issues that hinders the development of a more ecologically sound model. Profits in and of themselves are not 'good' unless they are ploughed back into the communities they serve. The acquisition of wealth for its own sake alone is an evil that has plagued humanity since the advent of farming.

The hoarding of wealth, the Dragon Hoards piled high by rich white men who contribute nothing meaningful to the world we actually live in, is nothing short of disgraceful. What do these mountains of money *really* add to life on Earth? Very little, if anything. Meanwhile, the rest of us, humans and non-human kin alike, are left to carry the cost.

We've forgotten what it means to have *enough*. The idea of sufficiency has been swallowed by the glittering bait of the celebrity cult. Everyone's chasing fame, chasing ease, dreaming of singing their way to millions without lifting a finger or touching the soil.

But it's time we snapped out of it. Time we grew up. The world doesn't need more idols or influencers—it needs people who are willing to get their hands dirty, to restore, to care, and to live as if what they do actually matters.

Another factor is *competition*: capitalism thrives on competition among individuals and businesses. Competing entities strive to offer better products, services, or prices to attract customers and gain market share. This competition is believed to enhance efficiency, innovation, and economic growth. However, this idea of competition is a lie; in late-stage capitalism, large companies own most of the smaller illusory companies. In Figure 1, you can see that only ten big companies own all of the actual food suppliers. This means that the Maccy cheese you buy from one supplier, which costs more money, and apparently tastes nicer because of that, is the same Maccy cheese that other companies sell for less profit. What you are tasting is the *placebo of luxury cost*.

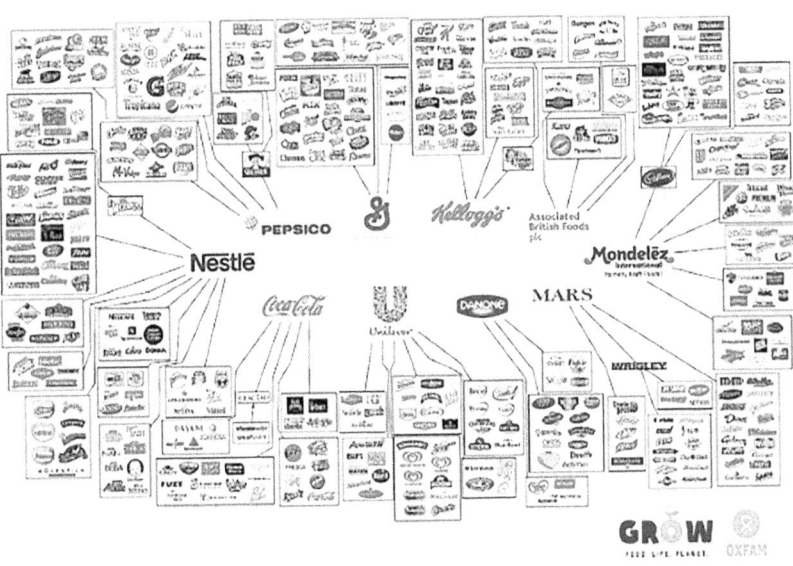

The lie of competition.[5]

---

[5] https://www-cdn.oxfam.org/s3fs-public/file_attachments/bp166-behind-the-brands-260213-en_2.pdf

Freedom of choice

Capitalism often presents itself as the champion of freedom, offering the freedom to choose, buy what you like, work where you want, and live your life on your own terms. But let us be honest. Freedom of choice, in this system, really just means freedom of consumption. You are free to choose from a range of products, free to pick your poison, but you are not free to opt out of the capitalist system. What choice do most of us actually have, other than to enter the endless grind of paid work? If you do not, you will have no house, no food, no warmth, and none of the so-called comforts that society dangles in front of you like a carrot on a stick.

Capitalism shouts about liberty while quietly binding us in chains. Yes, consumers can choose what to buy and from whom, and businesses can decide what to make and how to sell it. But this so-called choice is an illusion, and one of the most powerful tricks the system plays on us. We do not need shelves overflowing with options. We do not need 57 varieties of baked beans. What we need is two good-quality products and the time to enjoy them. All this excess, all this supposed abundance, just burns through our planet's resources faster, all to maintain the lie that we are free.

Real freedom of choice would mean being able to step away from this model altogether. It would mean having the right to live simply, to grow your own food, to build a life outside the economic machine without being judged as lazy, mad, or antisocial. But that kind of life is not celebrated. It is barely tolerated. Those who try are seen as dropouts or failures, while the rest of us are encouraged to keep running on the hamster wheel, working long days to pay for devices meant to save us time.

When you stop and really look at it, it makes no sense. We spend our lives working for others who profit from our labour, only to give that money straight back to them in the form of rent, bills, food, fuel, and all the trappings of so-called civilisation. And even when we do manage to buy a house or build a little something for ourselves, it is never truly ours. We will die, and what we leave behind is passed on—but even that is becoming harder, especially for our children, who are somehow worse off despite all the promises of progress. And gods help them if they have children of their own.

Yet if someone dares to say, "I do not want to live like this," they are met with suspicion, pity, or scorn. Society does not know how to deal with individuals who step outside the script. We are meant to conform, to consume, to keep the system alive.

And so I find myself asking, as many have before me, what is the alternative? There must be one. There has to be. But I do not claim to have the answers. Perhaps others, wiser and braver, will find the path. As for me, I cannot yet see a clear way out. Not one that most of humanity would even consider, let alone walk together. And so, for now, I try to live as gently as I can, to speak truth where I find it, and to hold on to the hope that another way might still be possible.

Population: The elephant in the room

One of the things that everyone seems to avoid discussing is the fact that there are simply too many of us on the planet, and our numbers are increasing. Population growth has a significant impact on the environment and ecosystems. As the population increases, there is a corresponding rise in demand for land, housing, agriculture, and the associated infrastructure. This leads to deforestation and biodiversity loss, threatening the survival of many plant and animal species. Loss of biodiversity can have a cascading effect on food webs in ecosystems, leading to the collapse of local populations.

As our global population continues to grow, so too does the demand for resources, clean water, energy, food, and raw materials. But the Earth is not limitless. The more we consume, the more pressure we place on already fragile ecosystems. We are seeing this everywhere. Fish stocks are vanishing under the weight of overfishing. Rivers are choked with runoff from fertilisers and intensive farming. Forests are stripped for timber, farmland, and profit, leaving behind scars where life once thrived. And all the while, the waste we generate, industrial, agricultural, and domestic, piles up, seeping into water, soil, and air.

This is not just about numbers or statistics. It is about people, places, and futures. As we burn more fossil fuels, cut down more trees, and continue with business as usual, greenhouse gas emissions rise, and so does the planet's temperature. We are now living through the consequences. Shifting rainfall, rising seas, heatwaves, floods, and droughts—climate change is no longer a distant threat. It is here, reshaping the world in real time.

And the natural world, already under strain, is struggling to keep pace. Habitats are shifting. Species are being forced to migrate or die.

The weather has turned fierce and unpredictable. It is not just nature that suffers; people suffer too, especially those already living close to the edge.

This is not about blame. It is about responsibility. About waking up to the reality that we are part of the Earth, not separate from it. We must find ways to live that nourish rather than drain, that restore rather than exploit. Because this isn't just about saving the planet. It is about saving ourselves.

One of the most significant 'con jobs' in Western culture is in the ignorance and wanton lies spread about the effects of climate change on our way of life and the lives of those who will inherit this Earth.

The rich, safely tucked away in air-conditioned towers and walled gardens, will barely feel the sting of climate collapse. And yet it's *them* who tell the rest of us, the ones scraping by as crops fail and homes flood, that it's all fine. Nothing to see here. Don't worry about the rising tides or burning forests, just keep up with the Kardashians and carry on consuming. It's classic distraction tactics: modern-day bread and circuses, spoon-fed to the masses while the Earth cracks beneath us. The empire burns, and we're meant to cheer for another celebrity wedding.

A solution must lead through sustainable practices and policies, as these are essential to mitigate the adverse effects of a growing population on our fragile ecology. This includes adopting renewable energy sources, practising responsible land use planning, promoting conservation efforts, implementing pollution control measures, and supporting sustainable agriculture and resource management.

Education and awareness are absolutely vital if we're serious about building a more sustainable future for both humanity and the Earth. However, let us be honest: this is not in the best interests of either industry or government. As someone who teaches within the UK education system, I can say with deep frustration and sorrow that we are not educating our children—we are conditioning them. We feed them a curriculum built on compliance, distraction, and celebrity culture, designed to maintain the status quo. We are raising not free thinkers, but obedient consumers.

And worse still, those children who *do* think differently—who see the world through another lens, who ask uncomfortable questions or process life in ways we do not fully understand—are labelled. We refer to them as children with 'special educational needs'. We mark them out as difficult when, in fact, many of them are simply seeing the world as it truly is. And far too often, our answer is to medicate them into compliance. Rather than embracing their perspective, we silence it with drugs, labels, and exclusion. These children could be our visionaries, our truth-tellers, our change-makers. But instead, we are flattening their spirit to fit the mould, and that mould is controlled by capitalism.

This suits the political establishment just fine, of course, as it keeps power in the hands of the wealthy elite and their corporate and banking allies. A compliant population is easy to control. An educated, questioning, passionate one is not.

Even if we manage to shift our thinking, and we must, we still face one of the most significant challenges of all: overpopulation.

The current global population is nearing eight billion. Around 169,000 babies are born every single day, while approximately 77,000 people die. That leaves us with a daily increase of around 92,000 human beings. On paper, this is simple arithmetic. In practice, it is a deeply tangled ethical and political crisis—one that strikes at the heart of our relationship with the Earth and with one another.

This comes down to women's rights. Because the only compassionate and truly effective way to address population growth is to ensure that women have the power to make free, informed choices about their own bodies. This includes access to contraception, to abortion, to complete reproductive healthcare without shame, restriction, or punishment. But instead, we are watching those rights being stripped away. Across the world, a wave of misogynistic, right-wing, theocratic ideology is dragging us backwards—reducing women to vessels, to objects, to state property once again.

If we are not vigilant, our bodies and our daughters' bodies will be pushed back through time, chained to outdated scripts of silence and service. This is not just a war on women. It is a war on freedom, on autonomy, on truth.

And here's the bitter truth: we are already seeing the Earth respond. Fertility is declining. Sperm counts have declined by more than 50% globally over the past fifty years. Rates of miscarriage and infant mortality are climbing in some areas, while many couples struggle simply to conceive at all. This is not a coincidence. This feels like a planet trying to regulate a species that has gone too far. The Earth is intelligent. She knows when the balance is broken and she seeks to restore it.

There is a deeper pattern here. The exploitation of the land and the oppression of women are not separate issues; they are entwined.

Throughout history, the bodies of women and the body of the Earth have been treated as resources to be owned, controlled, and consumed. Both have been silenced, commodified, and stripped of their sacredness. Where the land is ravaged, women's rights are often under attack. And where women are subjugated, the Earth is rarely treated with care. These are not coincidences—they are symptoms of the same worldview, one that values dominance over relationship, extraction over reverence.

And still, the hypocrisy reigns. Some men—and heartbreakingly, some women—claim to be pro-life. But they are not. They are pro-birth. Because if they genuinely cared about life, they would be feeding the children they force into the world. They would be lifting up mothers with real support, respect, and compassion. They would care about the woman's mind, her spirit, her whole being, not just her reproductive capacity. But they do not. Because in these circles, 'pro-life' often just means anti-choice, anti-woman, and deeply out of touch with reality.

And so, more and more women are choosing not to have children, not out of selfishness, but out of clarity. Because this world is poisoned. Because the cost is too high. Because it is not safe. Yet still, the pressure continues. The push for more births, more mouths, more control. No thought for the mothers, no thought for the future, and certainly no thought for the Earth.

It is a contradiction so vast, it feels almost ritual—a sacrifice offered on the altar of cognitive dissonance.

And yet ... There is still a flicker of hope.

Within Pagan and animist circles, I see something else. I see people speaking with care and honesty. People wrestling with complexity, refusing easy answers. People remembering that life is sacred, yes—but so is choice. So is truth. It is not perfect, but it is better. And that gives me a reason to keep going. A reason to believe that not all is lost. That some of us are still walking a path of sanity, justice, reverence, and care for the Earth, for each other, and for those who are yet to be born.

Ecological disaster

Sadly, the combination of an anthropocentric ideology put forward by those key monotheistic religions, along with the excess use of capitalism furthered by a lack of population control, has resulted in our presence on the Earth risking future sustainability.

Note—I am not saying that our actions will destroy the planetary ecosystem; they will not; our egos only believe that they will. We will eventually end or at least diminish in numbers to a sustainable level, and the Earth itself will continue. After all, the planet has had seven extinction events even before our appearance. She has an excellent immune system and can recover; we, on the other hand, will not be here to see it, at least not in the form we are now in.

As practitioners of a Green religion or philosophy, we must now actively participate in the ecological movement and try to make a difference in our local communities. We may not be able to influence world politics at the moment. Still, as individuals, we can practise good, ecologically sound livingness and start to make a difference in our sphere of influence.

Little steps make big changes if enough of us take the first one!

CHAPTER TWO

Paganism: The great green hope

To combat this almost nihilistic idea that we can simply use the Earth as we see fit with no consequence, we must first accept that the Earth is indeed a finite living being. Then, we can begin to consider the arguments presented by Animism and Spiritual Ecology. To initiate any restorative process, we must adopt an all-inclusive *ecological philosophy*, and only then can we restore the balance that we so desperately need to survive.

The word 'ecology' is derived from the Greek terms *oikos* (household) and *logos* (word or study). Thus, what we might term the study of the environmental house would include all its organisms and the functional processes that make the 'house' habitable. 'Ecology' can be taken further in that it promotes the study of 'life at home', emphasising more so the totality or pattern of relations between organisms and their environment. 'Ecology' encapsulates the philosophy of animism, as it acknowledges that the Earth is our home and that all beings have a place and a connection to us, and we to them.

So, what is Animism?
In Animism, everything in Nature, including plants, animals, rivers, mountains, rocks, and even natural phenomena like thunder and

lightning, possesses a conscious spirit or essence. These spirits are considered the source of life and energy and are often regarded as either ancestral or divine powers. It is one of the oldest known belief systems practised by indigenous and tribal societies throughout world history.

Graham Harvey, a professor of anthropology and scholar of religious studies, is known for his research on various aspects of indigenous religions, including animism. In his work, Harvey emphasises that animism should not be seen as a primitive or simplistic belief system, as it is often portrayed in Western analytical thought, but instead he argues that animism reflects a deep and sophisticated understanding of the world that is shaped by the cultural and ecological contexts in which it arises.

Harvey reminds us that animistic worldviews are grounded not just in a sense of interconnectedness but in a deep interdependence. Humans aren't set apart; we're part of a much wider community that includes animals, plants, rivers, stones, winds, and all manner of more-than-human beings. In these ways of seeing, non-human entities aren't just background scenery; they're persons in their own right, with agency and presence. This opens the door to relationships built on reciprocity, where responsibility flows both ways between us and the living world around us.

Harvey also stresses that animism isn't some abstract idea or dry intellectual theory; it's something lived, embodied, and experienced in the everyday. It appears in rituals, ceremonies, and quiet moments of reverence or connection. These acts aren't just symbolic; they're fundamental ways we engage with the spirit of the world. As Harvey puts it, these practices can have a genuine and powerful role in daily life.

One of his most important contributions, perhaps, is the call to break free from the old colonial lens. He challenges the tired, Western notion that animism is 'primitive' or naïve. Instead, he asks us to see it for what it is; a rich, sophisticated, and deeply rooted way of knowing, grounded in the cultural and ecological realities of Indigenous and local communities. It's not about looking back wistfully; it's about learning to listen differently.[6]

So, what we have is an excellent new or perhaps an 'old way' of viewing the world. A method that looks at creatures, plants, and animals as coexistent players on the planet, and whereby we can better

---

[6] Cf. Graham Harvey, *Listening People, Speaking Earth—Contemporary Pagansim*, 2nd edn (London: C. Hurst & Co., 2007).

accommodate and acknowledge the fact that non-humans have a greater or at least equal entitlement to the planet as do we humans. From this, I can now begin to unveil these *'entitlements'*:

The entitlement to life in and of itself, with no reference to what they are valued for by humans.

These include:

The right to live in a place of their own, without that right being superseded by human needs, e.g. housing or manufacturing.

The right to be treated humanely (and I know the irony of using that word when being used as food for humans).

Once individuals begin to understand these parameters, we can then start to remodel and realign any decisions made by the government, businesses, and indeed ourselves.

The movement from patriarchal, book-based religions to more individualised, earth-centred spiritual pathways, known as the so-called Dark Green religions, including Paganism and some forms of Christianity, holds out the hope of a more ecologically sound future.

'Dark Green Religion' is a term coined by Bron Taylor who uses it to describe a spiritual or religious orientation that strongly emphasises a set of beliefs and practises that view Nature not only as a resource to be exploited but also as a source of inspiration, value, and reverence. From this, Nature is therefore seen as sacred, and the interconnectedness of all life forms is recognised and honoured.

Green religions or philosophies often challenge the anthropocentric model that was central to more patriarchal religions, promoting an ecocentric perspective in which all life forms are equally valued and interconnected. These paths also inspire forms of environmental activism and advocacy, as they support adherents who feel a moral and spiritual duty to protect the environment and address specific ecological challenges.

Paganism is one of those Dark Green religious paths, and it presents a natural spirituality that attracts more and more people daily.

Defining Paganism is like trying to catch smoke with a butterfly net. Every time I think I've got it pinned down, I realise I've missed something, or said it in a way that doesn't quite land for someone else. Then along comes another practitioner with a slightly different angle, a tweak here, a nuance there, and I'm back to the drawing board.

But maybe that's not a weakness, perhaps it's our strength. Paganism doesn't demand a single, fixed doctrine handed down by some higher authority. Instead, it gives us the freedom to explore, to question, to investigate our own spiritual life on our own terms. And in doing so, we cultivate something the world sorely needs right now: a strong personal moral compass. One that's grounded not in fear or obedience, but in relationship with the Earth, with the divine, and with each other.

So, with that in mind, I'm going to try to be as clear and careful as I can, not because I think I've got the one true answer, but because I'd rather offer something respectful than stomp all over someone's deeply held path. Right then—deep breath—here goes:

Paganism refers to a broad category of spiritual and religious belief systems that typically fall outside those of the major organised religions (e.g. such as Christianity, Islam, Judaism, and Hinduism). The term *pagan* originated from the Latin word *paganus*, meaning 'rural' or 'rustic', and was initially used by early Christians to refer more specifically to people living in rural areas who practised traditional polytheistic religions.

Paganism encompasses a diverse range of belief systems and practices that are often rooted in ancient or indigenous traditions. These traditions may include reverence for Nature, such as polytheism (the belief in multiple deities) and animism (the belief that all objects and living beings possess a spiritual essence), as well as the worship of *ancestral spirits*.

Modern pagan movements often draw inspiration from pre-Christian European religions, such as Norse mythology (*Asatru*), Celtic and bardic traditions, as well as more obvious influences from Egyptian, Greek, and Roman polytheism. Alternatively, they may incorporate various elements of other indigenous spiritual practices from a variety of cultures worldwide. Some pagans focus on worshipping specific deities, while others emphasise the cycle of Nature, seasonal celebrations, and related rituals.

Paganism is a highly individualistic and decentralised belief system, where practitioners promote many beliefs and practices. There is no centralised authority or holy text that governs Paganism as a whole. Instead, pagans often engage in a more personalised exploration of their spiritual growth, where they can develop their rituals and practices based on a deeper understanding and connection to Nature and the Divine.

Q. How did I do?
The bit I struggle with, and I doubt I am alone in this, is that we have no central core from which we could reach out and influence the world.

The latest UK Census (2021) showed the following numbers for alternative religious views:

- Pagan (74,000)
- Wicca (13,000)
- Shamanism (8,000)

This equates to a total of 95,000 practitioners. That's a good-sized group.

This was the first Census that showed Christianity had declined to less than 50% of the population. The second most popular answer on the Census was 'no religion'.

If we could just come to a consensus on what we call ourselves, and I honestly don't care what that name is, but we could all use it, and from this, we would have a much larger representation in the public arena. One might ask—why do we need greater representation, certainly in a world where protesting and arguing against the current system is becoming dangerous? We need more people to develop a more 'mainstream' view to influence general political policies. We do not force others to our point of view, we do not perform terrorist acts in the name of our gods. We do not treat others with disrespect because of their gender, sexuality, or colour. Other paths do. Surely we need to share our own path to counter these ideas.

In paganism, there is an emphasis on the interdependence of all organisms and the importance of the natural world to both our mental health and our food resources. We can also acknowledge that the purpose of all life on the planet is not just a prerequisite to merely serving humans. We must begin to recognise that we are responsible to our children, our neighbours, and especially in our relationship to our 'green friends' and animal communities. So, before the total corruption and breakdown of our socio-economic and political institutions lead to the further deterioration of the environment, we, as pagans, must encourage others to become more incisive in playing a greater role in helping redress the balance that would otherwise ultimately lead to the destruction of our planet. Yet, we should ask—how can this be done?

Looking to our Elders

Indigenous peoples worldwide often have distinct perspectives and relationships with the Earth that differ from the concept of land ownership in more capitalist societies. Their ideas about land ownership are deeply rooted in their cultural, spiritual, and historical connections to the land.

Many indigenous cultures view themselves as custodians or caretakers of the land, rather than owners. The land is often regarded as a living entity, and indigenous peoples deeply respect its ecosystems, natural resources, and the interconnectivity between *all living beings*.

If we're serious about creating a more ecologically sound future, then we have to let go of the old colonial mindset, the one that assumes Western ways are always superior. It's time we started listening, really listening, to Indigenous peoples. Their knowledge, their relationship with the land, and their ways of living lightly and respectfully aren't just cultural artefacts; they're blueprints for survival. If we want to move forward, we need to do it together, side by side, learning from those who've been caretakers of the Earth far longer than industrial capitalism has been tearing it apart. It's not just about respect—it's about humility, and hope. Rob White states that:

> *Indigenous peoples have typically not separated cosmology from the economy. Their mythologies and religious practices were, and in some cases still are, integral to sustainable agricultural and ecosystem management strategies. As such, the economically driven destruction of the local habitats of indigenous communities is fundamentally a religious issue. To see it in any other way is to adopt Western capitalism's imperialistic, exploitative worldview, which is at the heart of the current ecological crisis. The globalisation of contemporary consumerist capitalism has placed indigenous societies under enormous pressure to assimilate into mainstream Western cultures and to make their lands available for the exploitatively destructive practices of forestry, mineral extraction, fishing, or tourism. According to this view, the destructive practices of Western-style development are simply normal and rational. Against this is a more sophisticated and less ideologically driven understanding that indigenous peoples have alternative development models that value homelands differently ... [that involve the] use of those lands [for] living for food, habitat, and trade. By so doing, they embody alternative models of sustainable life.*[7]

---

[7] Rob White, ed., *Controversies in Environmental Sociology* (New York: Cambridge University Press, 2004).

At the time of writing this book, I find myself deeply disappointed, though not surprised, by the re-election of Donald Trump. What is perhaps most disheartening is not only the direction of the United States under his renewed leadership, but the ripple effect it is having across the world. The fragile momentum that had been building around ecological restoration seems to be faltering. Green policies are being scaled back, dismissed, or outright reversed—not just in America, but in countries that once seemed committed to meaningful climate action. Instead of pushing forward, we are sliding backward. This comes at the worst possible time.

The United Nations had named 2020 to 2030 the 'Decade on Ecosystem Restoration'—a rallying call to heal what we have broken. The aim was ambitious but necessary: to restore damaged ecosystems, improve biodiversity, and regenerate the natural systems that sustain life. Restoration had moved beyond simply protecting endangered species or habitats. It had become central to maintaining human well-being, supporting agriculture, ensuring food security, improving water quality, promoting mental health, and enhancing the broader resilience of our communities.

This shift reflects the hard truth that we, as a species, have pushed the Earth beyond critical thresholds. The Planetary Boundaries model makes this clear: agriculture, land use, pollution, and resource extraction have driven changes that are now dangerously close to irreversible. Restoration is no longer a gentle, hopeful gesture—it is a lifeline. A way to keep socio-ecological systems functioning, to give future generations even a chance.

And yet, in the face of all this, political will is slipping. Corporate interests are tightening their grip. The vision of healing is giving way to short-term profit and nationalist posturing. What should be an age of renewal risks becoming an age of regression.

We *need* restoration, urgently, intelligently, and ethically. And we need it not just in wild places, but in the spaces where people live. That means recognising that restoration cannot be separated from the socio-cultural contexts in which it occurs. It is not just about planting trees or cleaning rivers. It's about changing mindsets, restoring relationships, and reweaving the threads that bind people to their place. It is about decolonising land management, listening to Indigenous knowledge, and reimagining what it means to live well with the Earth.

We were handed a decade to make a difference. And instead, in too many places, we are watching that opportunity be thrown away by those in power.

But not all hope is lost. Restoration is still happening—in pockets, in communities, in resistance movements, in sacred groves and gardens. And perhaps, in the end, that is where the real work has always lived, not in political slogans or summit declarations, but in the hands and hearts of those who are quietly refusing to give up.[8]

If we want to start living more ecologically and spiritually, we must acknowledge that we are not the only living beings on the planet with a 'right to live.' In returning to a more indigenous worldview, this is critical to our understanding of biodiversity and the ecosystem.

Q. What would a Green ethical system look like?
In her book *The Spiritual Dimension of Green Politics*, Charlene Spretnak describes the critical questions that we must ask to ensure that the values of the Green movement are understood and are therefore more workable. Allow me to paraphrase some of Spretnak's overviews:

Ecological wisdom

How do we operate our human societies in a more nature-centric way? How can we live within the planet's limits and build a better relationship between cities and the countryside? Can we think of ways to guarantee the rights of non-human species?

Grassroots democracy

How can we encourage and develop new systems that enable us to control the decisions that affect our lives, allowing mechanisms in planning and development that align with the community's focus and take into consideration the impact these decisions have on future generations?

---

[8] Cf. Carter A. Hunt and others, 'Setting up Roots: Opportunities for Biocultural Restoration in Recently Inhabited Settings', *Sustainability*, 15 (2023), 2775.

Personal and social responsibility

How can communities and governments respond to human suffering that promotes human dignity? How can we encourage people to make healthier lifestyle choices and promote community health? How can we develop a community education system that supports learning for its own sake, rather than serving the machine of industrial capitalism?

Non-violence

How can we develop an alternative to the current violence which is prevalent within all communities? How can we eliminate weapons of mass destruction, such as the nuclear arsenal, without displaying naïveté towards other governments with ill intent?

Decentralisation

How can we restore power and responsibility to individuals while encouraging communities to work together on the smallest scale, promoting community-based economics rather than the current trend towards globalisation, which causes massive ecological damage through the worldwide transport of goods and materials?

Reevaluating personal identity

How can we move away from the narrow job ethic or identification with our professions? And what will be the new definitions of income, wealth, and work? We must be able to move towards the idea that a country's GDP should be based on the health and well-being of its society, how it cares for its vulnerable members, and the standard of living for the community's poorest members. The growth of a country should not be based on financial profit for the few; instead, it should be focused on the quality of life for the many.

Respect for others

How can we change our culture to honour cultural, racial, sexual, religious and spiritual diversity? How can we move from the climate of intolerance and xenophobia that is currently gripping many countries?

Global responsibility

How can we have genuine programmes of levelling up for third-world countries—sharing technology freely, giving medical and financial aid without expectation of repayment? How do we provide education on preventing pregnancy and population growth without judgement or trying to impose religious limitations?[9]

What's impressive is that, across the globe, a powerful and positive shift is underway. A growing number of countries are moving beyond GDP as the sole measure of success and embracing more holistic indicators of well-being, sustainability, and equity. Scotland has recently joined nations such as New Zealand, Iceland, Wales, Finland, and Canada in forming the Wellbeing Economy Governments partnership, which champions economic systems designed to serve both people and the planet, not just profit. New Zealand has pioneered a Living Standards Framework, Bhutan has long led with its Gross National Happiness index, and Canadian provinces have adopted a National Index of Wellbeing to track quality of life across multiple dimensions.

Even in the United States, states like Hawaii, Vermont, and Maryland use the Genuine Progress Indicator to assess environmental and social impacts alongside economic performance. These efforts are reinforced by international bodies such as the OECD and the United Nations, who increasingly advocate for a wellbeing approach to policy. This marks a hopeful departure from growth-at-all-costs thinking—a global recognition that a truly prosperous society is one where people thrive, ecosystems are protected, and future generations have a chance to flourish.

If we can move away from profit-centred, self-interested economic models currently being used, we may be able to create a more successful society for all, including both humans and non-humans.

In her contribution 'The Greening of the Self' in the book *Spiritual Ecology: The Cry of the Earth*, Joanna Macy discusses transitioning from what she calls the Industrial Age to the Ecological Age. I would like to share a few of her insights.

Macy views the Industrial Age as Mechanomorphic, with the universe seen as a machine. She sees that in this age, Earth would be viewed

---

[9] Cf. C. Spretnak, *The Spiritual Dimension of Green Politics* (Santa Fe, NM: Bear Co., 1987).

as inert matter consisting of atoms, with life seen as random chemistry, rather than the view she takes in the ecological age, where the universe is Orga*nic*, seen as an ongoing process or story. Gaia is acknowledged as a Superorganism. The holistic, autopoietic worldview is the only perspective from which to view this.

Macy goes on to describe the role of the human in these two systems. In the industrial age, humanity's role is often seen as one of conquering nature, which is frequently perceived as degraded; this allows the anthropocentric view to dominate, and thus, nature has no instrumental value, but is instead viewed as almost background noise.

In the Ecological age, however, humans are seen as stewards in the proper sense, restoring the natural balance and living in symbiosis with the natural world. Nature has intrinsic value, and this view enables social ecology and ecological justice.

The road to making these changes will be long and hard, but with personal changes in our own world, we can go a long way to making these things happen. However, we also need to raise our voices in the fight against multinational corporations, which have caused most of the damage to our society and ecology, as well as combating the apathy and acceptance of many in our society towards the way that we live and the fact that many of us are guilty of not living ecologically or sustainably.[10]

---

[10] Joanna Macy, 'The Greening of the Self', in *Spiritual Ecology: The Cry of the Earth*, ed. Llewellyn Vaughan-Lee (Point Reyes, CA: The Golden Sufi Center, 2013), pp. 145–58.

CHAPTER THREE

An Oxymoron: 'Ecologically Unsound Paganism'?

After speaking so passionately about the Pagan way of life and recognising that animism offers one of the clearest and most powerful responses to the ecological crisis we face, it might feel jarring to now turn the mirror on ourselves. But we must. Despite all our talk of connection, reverence, and resistance to the dominant system, many of us are still slipping back into old patterns of commercialism. We are handing over our spiritual and economic agency to big business—yes, even in our magic.

What do I mean by that? Well, brace yourself.

Let's talk about the uncomfortable bits. Let's face them head-on, not to shame ourselves, but to grow. Somewhere along the way, many of us have fallen into the trap, some out of innocence, others out of unawareness, of treating our magical practice as something that needs to be bought. Our altars are filling up with imported trinkets, mass-produced tools, and branded 'witch kits' that promise instant power. We scroll, click, and purchase, chasing a sense of authenticity that we already carry within us.

But here's the truth, it matters *how* we practise. It matters *what* we use and *where* it comes from. Because magic rooted in animism cannot be divorced from the Earth or from the impact of our choices. So let us now take a closer look, not to scold, but to ask: does this serve the magic, or does it serve the machine?

Because the path we walk is sacred—and we must walk it with our eyes open.

The use of crystals in magic and Paganism

Crystals have become almost synonymous with modern Paganism. We wear them, we work with them, we place them on our altars, and we speak of their energies with reverence. In some circles, there is even an unspoken pride in the sheer number of crystals we own, as if the weight of our collection somehow reflects the depth of our spiritual path. But here's the truth we often avoid: this is not just a personal practice—it's part of a billion-pound global industry. One that thrives on extraction, exploitation, and environmental devastation.

We cannot continue to ignore this. So let's talk, really talk, about what it means to use crystals in a world crying out for healing.

Crystals do not gently rise from the Earth, ready for us to pluck them like apples from a tree. They are mined, ripped, blasted, and clawed from their ancient homes through processes that are anything but gentle. Whether it's open-pit mining that scrapes away whole ecosystems, or deep-shaft operations that scar the Earth and endanger the lives of workers, some of whom are children, the truth is the same: the crystal on your shelf came at a cost.

Geologists survey the land, often in remote and vulnerable regions. The topsoil, home to plants, insects, and countless unseen life, is stripped away with heavy machinery. Blasting and drilling follow, sometimes using explosives, to break through the rock and reach the crystal veins. Entire landscapes are reshaped. Rivers are diverted. Forests are cleared. Indigenous lands are violated. All so that we, in the comfort of our homes, can hold a polished stone in our hand and feel 'spiritual'.

And after all that destruction, these crystals are cleaned, sorted, shaped, and shipped halfway across the globe—wrapped in plastic, packaged in glossy boxes, and sold to us as sacred tools. This is not magic. This is consumerism dressed in mystical robes.

Here's where the contradiction becomes unbearable: many of us who walk Pagan and animist paths deeply *care* about the Earth. We speak of connection, of reciprocity, of reverence. We talk of spirits in stone and the wisdom of the land. And yet, we continue to buy into an industry that is built on violence against that very Earth.

Some try to ease this discomfort by seeking out so-called 'ethically sourced' crystals. And I understand that desire—we want to believe we are making better choices. But a recent research paper has made something painfully clear: these ethical labels are, for the most part, a fiction. Investigators found that suppliers using these claims were unable to offer any concrete evidence to support them: no traceable mining practices, no verified labour conditions, no genuine oversight. Just pretty words on packaging designed to soothe our conscience and keep the money flowing.

Let me be blunt: we are being lied to. We are being taken in. And it is time to wake up.

Now, please do not panic. I am not asking you to cast out the crystals you already hold dear. Many of us have deep, powerful relationships with the crystalline realm. These beings have offered healing, clarity, and connection. That work is real. That relationship *matters*.

But what I *am* asking, what I am urging, is that we stop buying more. That we say no to this destructive cycle. That we honour the crystals we already have, treat them as sacred allies, not consumables, and choose to end our role in the system that is ripping them from the Earth.

This industry runs on demand. If we withdraw our financial support, it will begin to falter. If we educate ourselves and others, it will lose its grip. If we realign our practice with our values, we will begin to reclaim the heart of our magic.

Let us go back to the land. Let us work with the stones beneath our feet, the ones still nestled in their proper homes. Flint from the field.

Pebbles from the stream. Stones from ancestral paths and local quarries. Ask permission. Make offerings. Build a relationship. That is animism. That is Paganism.

Because the path we walk is not built on glittering shelves and branded boxes. It is built on dirt under the nails, on honouring what lives, on fierce love for this sacred world.

So hold your crystals close. Speak to them. Learn from them. But let that be enough. Let this be a line we choose not to cross. For their sake. For the Earth's sake. For our own integrity as spiritual people.

The extinction of sacred plants

The second thing we urgently need to examine is our increasing use of sacred plants within magical and spiritual practice. It is no longer enough to speak of reverence while unconsciously participating in patterns of overuse, extraction, and environmental harm. Many of the herbs, oils, and plant materials we burn, carry, distil, and blend into our ritual work are now under threat. Some are facing extinction, not because they have been forgotten, but because they have been *over-loved by an ever-growing pagan population*, consumed beyond their capacity to regenerate.

In our pursuit of spiritual connection, we have, at times, become consumers first and practitioners second. Our incense bowls overflow. Our cupboards are filled with imported oils and powders. And yet few of us stop to ask where these plants came from, how they were grown, how they were harvested, or who paid the price so that we could feel more magical.

This is not just an ecological issue, it is a spiritual one. If we claim to walk an animist path, to see the world as alive, then we must recognise that each plant we use has a spirit. A life. A history. And when we burn or brew that plant without relationship, without consent, without gratitude or offering, we are not practising magic, we are enacting control.

Again, as seen with crystals, labels like 'sustainably harvested' or 'ethically sourced' sound comforting, but without transparency, they are often little more than greenwashing—designed to soothe our conscience while maintaining profit margins. Many Pagans and witches are being misled by these claims, lulled into a false sense of integrity by branding that has no roots in verifiable truth.

This is not a call to shame—but it *is* a call to awaken.

We must slow down. We must stop reaching for more. We must begin to reimagine what our practices could look like if they were deeply rooted in relationship, locality, and respect. There is powerful magic in working with the plants that grow around you—in forming a bond with the wild herbs of your own land, the green ones who know your weather, your soil, your breath. This is not a lesser magic. It is a deeper one.

So let us commit to learning. Let us make offerings. Let us ask permission. Let us be willing to say, "No, I will not use this plant, not now, not like this." Let us remember that true connection cannot be bought in a packet, nor authenticity measured by the contents of a ritual cupboard.

This is not about austerity. It is about integrity.

To be in right relationship with the plant spirits is not to collect them—it is to walk beside them, to listen, and to act in ways that honour the living Earth, not just our spiritual ego.

If our practice is to mean anything, it must reflect the values we claim to hold. So let us choose a path of wisdom, of restraint, of reciprocity. The plants are listening. The Earth is watching. And the choice is ours.

Let us discuss some of the plants that are in danger.

Frankincense

Frankincense is a resin obtained from trees of the Boswellia genus, primarily *Boswellia sacra* and *Boswellia carterii*. Its aromatic properties and use in various cultural, religious, and medicinal practices have been highly valued for centuries.

The trees that produce frankincense are native to regions of the Arabian Peninsula, specifically Oman, Yemen, and Somalia, as well as parts of North Africa and India. The resin is obtained by making small incisions in the tree's bark, allowing the sap to slowly ooze out and harden into solid, tear-shaped droplets.

As many of you will be aware, frankincense has a beautiful fragrance when burned. It has been historically used in religious ceremonies for centuries, particularly in ancient civilisations such as those in Egypt, Mesopotamia, and the Mediterranean region. It is burned as incense, then offered as a sacred gift to deities, or it can be used to purify and sanctify a holy space. When the ancients used frankincense in their rituals, its use was fairly minimal. However, the use of frankincense today by pagans, the Catholic and high churches, and individuals has subsequently grown exponentially.

If you consider pagans as a group, referring back to the UK Census, there are nearly 100,000 of us, and I can almost guarantee that we all have frankincense in the house. Can you now see where I am going here? The demand for the blood of this sacred tree is massively outstripping the ability to supply. We are literally bleeding the trees to death. Add to that the effect of climate change, and you will see that there are massive implications for this plant's survival.[11]

---

[11] Frans Bongers and others, 'Frankincense in Peril', *Nature Sustainability*, 2 (2019), 602–10.

White sage

Once again, we are confronted with a stark example of the deeply ingrained anthropocentric mindset, the belief that plants, like everything else in Nature, exist solely for human use. It's a mindset that bulldozes through ecosystems and cultures alike, treating sacred life as product, and heritage as branding. Nowhere is this more painfully clear than in the commercialisation of white sage (*Salvia apiana*).

White sage is a sacred plant that has been used for generations in purification, prayer, and healing ceremonies by Indigenous peoples of southern California and Baja California, particularly by the Chumash, Tongva, Cahuilla, and other Nations. It holds profound spiritual and cultural significance; its smoke is not simply a tool, but a prayer in itself, a relationship, an offering.

And yet, in recent years, this plant has been swept into the global marketplace under vague banners of 'wellness', 'spiritual cleansing', and 'good vibes'. Smudge sticks—many of them wrapped in glitter, ribbon, or labelled with appropriative terms—are now sold in supermarkets, online boutiques, and yoga studios, often with no mention of the cultures from which they were taken and with no consideration for the harm this causes.

Make no mistake, this is not just cultural appropriation, though it absolutely is. It is also an ecological crisis.

White sage is a slow-growing perennial that thrives only in specific arid regions of southern California and northern Mexico. These ecosystems are already under strain from prolonged drought, climate change, and increasingly frequent wildfires. The rising demand for white sage has led to illegal and unethical harvesting on public lands, in national parks, and on protected Indigenous territory, often without permission and without any sustainable practices. In many cases, entire plants are being uprooted, rather than cut back responsibly, which halts their ability to regenerate and destroys the surrounding habitat.

As a result, conservationists and Indigenous communities alike have raised serious concerns. While white sage is not yet officially listed as endangered, it is categorised as a species of concern, and wild populations are showing marked signs of decline. Localised extirpation (disappearance in certain areas) has already occurred.

So we must ask: when sacred plants become trendy commodities, who truly benefits? And who bears the cost?

It certainly isn't the Instagram influencer waving a smudge stick between yoga poses. It isn't the online shop selling mass-produced bundles to 'cleanse bad energy'. The real cost is borne by the land that is being stripped bare, by the plant whose spirit is being commodified, and by the Indigenous peoples who have spent generations protecting it, only to see it exploited in the name of a spirituality that erases them.

That is the cost of unchecked spiritual consumerism. And we have to decide whether we're willing to keep paying it.

The fate of the teacher plants

This brings me to the teacher plants, the ones that walk with us in the liminal spaces, the ones who crack us open and put us back together, if we let them. These are not just herbs or hallucinogens or curiosities to be categorised. These are ancient beings. They have taught us for thousands of years, served as guides, healers, challengers, and allies. They are bridges between the worlds, gatekeepers of the sacred, mirrors held up to our souls.

But in our relentless human way, we have not treated them with the care they deserve. We haven't simply asked to learn, we've demanded. We've extracted. We've turned teachers into tools, medicines into merchandise, and sacred encounters into products to be sold, marketed, and consumed. And of course, there are consequences. There are always consequences when reverence is replaced with entitlement.

Ayahuasca

Ayahuasca is not a trend. It is a sacred medicine, a powerful psychoactive brew made from the *Banisteriopsis caapi* vine and the *Psychotria viridis* leaf. For generations, stretching back beyond memory, it has been used by Indigenous communities of the Amazon for healing, visioning, and deep spiritual communion. This brew is not taken lightly. It is part of a living lineage, woven into the heart of ceremony, community, and relationship with the rainforest itself.

And yet, in recent years, ayahuasca has been swept into the growing tide of global curiosity; repackaged for seekers, psychonauts, and spiritual tourists chasing revelation. Thousands now travel to places like Peru, Brazil, Ecuador, and Colombia, eager to sit in ceremony. Some are guided with genuine care by Indigenous healers. Others find themselves in ceremonies led by non-Indigenous facilitators whose training, understanding, and cultural grounding vary wildly.

This surge in demand is not without consequence. The plants that comprise ayahuasca are being increasingly overharvested. The vines take years, sometimes decades, to reach full potency. In many regions, they are being stripped from the forest faster than they can regrow. As with so many sacred plants, the rush to consume them is placing enormous strain on delicate ecosystems and threatening the long-term health of the very plants people claim to honour.

But it is not just the land that pays the price.

As ayahuasca is pulled into the global market, its meaning begins to fragment. In some spaces, ceremony becomes performance, tailored to the expectations of tourists rather than rooted in the slow, respectful traditions it comes from. In more disturbing turns, it has been co-opted entirely. Some Western companies are now sending employees on 'spiritual retreats' involving ayahuasca, marketed as wellness initiatives or team-building exercises. These sessions are often stripped of any cultural context or spiritual integrity, reduced to exotic experiences for corporate gain.

At what point do we stop and ask, where is the spirit in all of this? Where is the relationship? The reverence?

There's no shame in seeking healing. There's no fault in wanting transformation. However, when our pursuit of spiritual depth comes at the expense of Indigenous sovereignty, fragile ecosystems, and the plants themselves, then we must pause. We must listen. And we must ask: is this still medicine, or has it become just another form of consumption?

The plants deserve more than to be treated as commodities. So do the people who have carried their songs, their stories, and their teachings through generations of colonial violence and erasure. We do not need to turn away from ayahuasca entirely. But we do need to turn towards it with humility. With respect. With slowness. And with the willingness to hear no.

This is not just about ethics—it's about relationship. And a genuine relationship takes time. It takes listening. It takes responsibility. The forest is still speaking. The question is, are we willing to stop speaking over it long enough to hear?

Tobacco

For many Indigenous cultures across the Americas, tobacco—known by names such as *Asemaa*, *Cansasa*, and *Kinnikinnick*—is not just a plant. It is a sacred relative. It is offered in prayer, carried in bundles, placed on the Earth, or burned with intention. It is a messenger plant, used to carry words to the spirits. It is called upon for protection, for grounding, for clarity. It opens the lungs and the blood, sharpens the senses, and clears the channels between the physical and the unseen.

When used respectfully, in ceremony, with intention and humility, tobacco is a powerful ally. A proper medicine.

But then came colonisation and industrialisation. Stripped of its ceremonial roots, tobacco was twisted into something else entirely. It was commodified, mass-produced, genetically modified, pumped with chemical additives, and marketed as a habit. It became recreation. It became an addiction. It became death.

And in that process, the spirit of tobacco was desecrated.

It is my personal belief that the suffering caused by tobacco today is not just a physical issue. The cancers, the respiratory failure, the widespread addiction; these are not merely side effects. They are spiritual symptoms. Warnings. Echoes of imbalance. The spirit of tobacco, once honoured, has been dishonoured. Its role has been ignored, its power misused, and its voice silenced.

This isn't punishment. It's not wrath. It is a call back to right relationship. The disease is the message: *You have forgotten the sacred. You have taken without giving. You have stopped listening.*

And we're making the same mistake again.

Cannabis

Once a revered plant spirit, cannabis was known for its gentle, calming nature. It was used ritually across cultures—for visioning, for rest, for pain relief, for quiet communion with the sacred. It whispered to the soul. It softened the edges of pain. It offered spaciousness and insight when used with care, presence, and purpose.

But now? That sacred whisper has become a scream.

In our hunger for stronger highs, cannabis has been selectively bred and modified into increasingly potent strains, engineered for maximum psychoactive effect. THC levels have skyrocketed. The balance that once made cannabis such a powerful ally has been disrupted, torn apart by market demand and laboratory precision.

And unsurprisingly, this shift has consequences. The dramatic rise in strength is now directly linked to increasing rates of psychosis, particularly in young people. This is no longer the ancient plant of soft wisdom and mellow visioning. It has become a distorted echo of itself, loud, overwhelming, and often damaging to the very minds it once helped soothe.

Where cannabis was once honoured, it is now consumed relentlessly. Stripped of context. Stripped of spirit. It has become a product, a trend, a coping mechanism, a political pawn.

And speaking of politics, let us not ignore the grotesque irony at play here.

Cannabis remains a Class B drug in the United Kingdom. Possession can lead to arrest, prosecution, and even imprisonment. Unless, of course, you happen to be married to power. The largest medicinal cannabis-growing company in the UK is part-owned by the husband of former Prime Minister Theresa May. Yes, while young people are criminalised for carrying a small amount of cannabis in their pocket, the powerful profit from its large-scale cultivation and export under the banner of curaleaf. This is hypocrisy at the highest level.

As long as you pay for it, it's legitimate. As long as it's backed by investment portfolios and government deals, it's legal. However, when it is cultivated by communities or used by individuals for spiritual or personal healing, it becomes a criminal act.

So we must ask: how did a sacred plant become both a controlled substance and a corporate asset? How did something once associated with connection, peace, and insight become twisted into a weapon of punishment for the poor and profit for the rich?

This is the danger of taking without asking. Of consuming without reverence. Of demanding a spiritual experience without being willing to carry the responsibility that comes with it.

Teacher plants are not here to serve our ego. They are not here to help us bypass pain or amplify illusion. They are here to transform us, but only if we come with integrity, with care, and with listening hearts.

And if we cannot come that way, if we cannot honour the plant, the land, the culture, and the spirit, then perhaps we are not ready to receive their medicine at all.

And then there's this.

What the fuck have you done to our food?

Once, wheat was our companion. We walked with it for thousands of years, shaping it gently, letting it shape us in return. But that old covenant, seed, soil, and patience, has been broken. Wheat is no longer the same living being it was. It has been rewritten, gene by gene, in the name of 'efficiency'.

Through gene editing and transgenesis, corporations are driving wheat to grow faster, resist drought, and push grain production beyond its natural limits—in pursuit of the goal of raising global yields from today's 3.3 tonnes per hectare to 5 tonnes per hectare by 2050.[12] CRISPR-edited wheat, promising a 10% yield increase, is already being trialled in Australia, with a commercial release expected within a few years.[13] It is no longer a plant in relationship with its land; it is a commodity engineered for the numbers on a spreadsheet.

And yet, for all this 'progress', the food we have trusted for millennia is now turning against us. Up to 20% of the population is now intolerant or sensitive to FODMAP proteins, including those found in wheat, which our guts once digested with ease.[14] Our own bodies are rejecting what has been done to this food.

But instead of taking responsibility, the same corporations that broke the wheat now tell us it is our fault. We are 'weak'. We are 'allergic'. And so, we pay again—handing over our money to the booming gluten-free and wheat-alternative industries that are owned by many of the very same companies that engineered the problem. We are literally buying back our right to eat safely, paying for the privilege of being poisoned by this Frankenstein grain.

And what of the land? Every genetic tweak alters not just the plant but its place in the web of life. Engineered wheat alters the sugars exuded by its roots, changing the soil microbiome, damaging fungal networks, and contributing to the silent demise of the living earth beneath our feet. Yet soil health, biodiversity, and even basic human digestion are ignored—all sacrificed to yield and profit.

All of this points to a deeper truth. If we continue to treat plants as mere resources to extract, modify, and package, then we will continue to suffer from health issues. Not just in body, but in spirit. These plants, whether psychoactive or mundane, are not ours to exploit. They are beings with wisdom and agency, and we ignore that at our peril.

---

[12] Nikolai Borisjuk and others, 'Genetic Modification for Wheat Improvement: From Transgenesis to Genome Editing', *BioMed Research International* (2019), 6216304.
[13] R. S. Stokstad, 'Australian Trial of Gene-Edited Wheat Aims for 10% Yield Increase' (2024) <https://www.reuters.com/markets/commodities/australian-trial-gene-edited-wheat-aims-10-bigger-yields-2024-05-23>.
[14] Caroline J. Tuck and others, 'Food Intolerances', *Nutrients*, 11(7) (2019), 1684.

When we disrespect the teacher plants, when we remove them from ceremony and context, when we put them on factory belts and sell them by the kilo, they *change*. Their energy changes. Their impact on us changes. And not for the better.

We need these plants, but not as products. We need them as guides. And to hear their voices, we must return to humility, to reverence, and to right relationship. That means ritual. That means asking permission. That means defending the lands they grow in and respecting the people who've carried their knowledge for centuries.

In Part 2, I'll share some practical alternatives for those of us practising here in the UK. These are local, context-based allies, plants, materials, and methods you can find right on your doorstep. Working with them not only deepens your connection to the land you actually live on, it also helps you step away from the endless pull of consumer culture.

Candles and ritual leavings

The use of light in ritual is as old as time. Across cultures and centuries, we've used oil lamps, tallow candles, and hearth fires to symbolise spirit, presence, and prayer. In my own coven, we try to make our practice more ecologically sound by using beeswax when possible. However, the truth is that most of us use tea lights.

They're practical. Small, self-contained, easy to use, and with a finite burn time, they are perfect for casting circles or marking sacred space. But here's the problem: tea lights are made from paraffin wax, a by-product of oil manufacturing. That's already an ecological red flag. Add to that the aluminium or tin casings, and suddenly your 'harmless' little flame starts to look a lot less innocent.

Worse still, many of these candles are left behind at sacred sites, places of pilgrimage, ancient stones, and woodland altars. People leave them to honour a deity, a spirit, or a lost loved one. The intention is good. But the impact can be devastating. Birds mistake the wax for food and are poisoned. The metal casings don't break down, they remain lodged in the Earth, another relic of our thoughtlessness.

The simplest solution is this: let your candle burn out or cool completely, then take it home. Place it in your recycling bin or landfill—don't leave it in the wild. Ritual does not end when the flame goes out. It ends when we have taken responsibility for the space we have worked in.

The same applies to other so-called offerings, such as coins, salt, and incense wrappers. Coins, in particular, are harmful to trees. What many don't realise is that hammering coins into a tree 'for luck' or 'as an offering' can, over time, kill it. Copper, especially, is toxic to many tree species. In fact, an old (and quite grim) method of circumventing tree protection orders was to drive copper nails into trees. Once the tree died, it could be removed, legally. Imagine that. Poison as a loophole.

We often do these things, unknowingly, as part of our wish-fulfilment, hoping for love, healing, a sign, or a blessing. At its core, however, this is a form of anthropocentrism. We are placing our own needs above the well-being of the land, the tree, and the bird. We dress it in sacred language, but it's still extractive. Still one-sided.

If our ritual actions cause harm to the very beings we claim to honour, then it is not magic. It is a contradiction.

We need to rethink our offerings. Reframe what it means to give. A song, a handful of seeds, a breath, a prayer, a clearing of litter—these are gifts the Earth can receive without suffering. These are offerings made in *right relationship*.

Let our witchcraft be love in action—not just a ceremony of desire, but a promise to tread lightly, and to give back more than we take.

Clooties

We have all seen a variety of *Clooties* tied to trees, an ancient way of making magic. It is a time-release type of spell for good or ill. As the cloth rots away, the magic is then released over time. Many individuals use clooties as a remembrance for lost loved ones, and also as a mark of having been to a particular sacred site where a magical ritual may have been performed.

However, many who are perhaps less aware are using non-biodegradable materials in the clootie and here the problem begins. I have seen nylon tights, florist ribbons, car chains and even knickers (though I am not sure what they were asking for) tied around the branches of trees. Anything that does not rot or is tied too tightly around the tree will cause the tree to start growing around it, cutting into the xylem tissue.

The clootie will essentially act as a tourniquet. Thus, like tourniquets that are tied around an arm or leg which is bleeding, the clootie can damage the tree! And as environmentally aware pagans, we should avoid this practice at all costs. If you want to use clooties magically, use hemp, paper, or another substance that can disintegrate quickly. This practice releases your magic as it should and saves the tree pain and anguish. There is also the impact on local wildlife to consider.

This 'little man' (below) did not know the difference between clooties and leaves and had then eaten one, which then got stuck in his throat. No doubt a hefty bill for the farmer here, and if he cannot afford to pay it and sees the calf as 'not worth the spend', this young man's life will end. Yet again, through a simple lack of understanding, we often think only of ourselves, and through these thoughtless acts, we fail to consider the environment we claim to promote.

Stone piling

Stone piling, also known as cairn building or rock stacking, refers to stacking rocks or stones on top of one another in or near rivers, streams, or other natural water bodies. While it may seem innocuous and serve as an artistic expression, it can harm the environment. Building stone piles in rivers can disrupt and damage the natural habitat of aquatic organisms. It alters the physical structure of the riverbed, potentially destroying critical microhabitats for fish, invertebrates, and other aquatic species. The arrangement of rocks can disrupt the habitats of inhabitants, including their feeding, breeding, and nesting grounds, thereby affecting their population dynamics and the overall health of the ecosystem. It also alters the natural water flow pattern and increases erosion in river channels.

Moving rocks can also crush or damage eggs, larvae, or small invertebrates that rely on the spaces between rocks for shelter and protection.

Stone piling might look harmless, even 'artistic' to some, but it's yet another example of humans stamping their will onto the natural world. It's that old arrogance again—us first, the land second—as if the Earth exists to bear our marks. Honestly, the land is far more beautiful left to its own wildness; we don't need to decorate it with our little rock sculptures. We've already littered the world with enough phallic buildings in our cities: do we really need to drag that energy into the woods as well?

The wounded waters

Water is the first mother. Every culture, every sacred text whispers it. In the Bible, creation begins with a formless deep, with the breath of God hovering over the face of the waters. Science says the same in different words—life crawled out of those primordial seas, and we are still, quite literally, walking bags of ancient water. It moves through us as blood, sweat, and tears; it keeps us alive. Without it, we die in three days. But worse than dying of thirst is what we are doing now: we are killing the very thing that birthed us.

For Pagans, animists, and anyone who listens to the wild world, water is not just a resource. It is kin. We kneel at healing wells, we sing to rivers, we whisper prayers to springs where spirits dwell. To stand in cold running water, to feel it rush around your legs, is to feel the pulse of the world itself. Rivers are alive. Wells have voices. Streams carry stories. To work magic with water is to work magic with life itself.

And now? Now our waters are crying out. Sewage, chemicals, and plastic choke them. Raw effluent pours into rivers once sacred to gods and spirits. The sanctity of those places is being violated. The spirits feel it; *we* feel it. Those of us who work with wild waters are not immune, we carry their wounds in our own hearts. The river that once sang now feels heavy, dull, its energy sluggish and broken. To step into polluted water is to feel grief rise in you like a tide.

Eco-anxiety is no longer an abstract fear of 'the climate crisis'. It is here, visceral, in every Pagan who lays offerings by a stream, in every wild swimmer who feels their skin prickle with unease in filthy water. The bond we have with water makes its suffering our own. To watch it darken and sicken is like watching a beloved friend being poisoned in slow motion and knowing we are complicit simply by living in a world that allows it.

The worst part? We take it for granted. We act as though we can dump, poison, and drain, and the water will forgive us because it always has. But there is a breaking point. Polluted rivers are not just ecological disasters; they are spiritual wounds. When we kill the water, we are killing something sacred, something that has always loved us enough to keep us alive.

And maybe that's the truth we don't want to face: if water is the mother of life, then we are guilty of matricide.

The problem with poop and profit

Our rivers are dying, and the people paid to protect them are the ones poisoning them. Water companies are dumping untreated sewage into rivers as if it were their right, choking them with algal blooms, fatbergs, and chemical sludge. Fish die, wildlife vanishes, and not a single river in the UK can now be called clean.

And why? Because profit comes before water.

Thames Water is a prime example, bankrupt and begging for a government bailout, while quietly paying shareholders almost the same amount they owe in debt. It's not an accident; it's a business model. Across the UK, water is no longer a sacred responsibility; it's a cash cow. Sewage in rivers? A minor inconvenience compared to the duty to keep the money flowing to investors.

The government's own 2022 report admits it: sewage dumping, fertiliser run-off, plastic waste, and chemicals are creating dead zones and breeding grounds for antimicrobial resistance. Yet the outrage is left to groups like Surfers Against Sewage, wild swimmers, and eco-activists. At the same time, the companies shrug and carry on, treating her like a landfill for the sake of shareholder payouts. This isn't just mismanagement; it's sacrilege.

Chemical pollution

Chemical pollution disrupts aquatic ecosystems by poisoning organisms and altering their reproductive cycles, including those of humans. The level of oestrogen in London's water system is slowly but surely causing a decline in sperm motility and fertility rates in men. Medicines that cannot be filtered out of the water also contribute to a weakened immune system response and the development of antibiotic resistance. During the COVID-19 pandemic, there was evidence of high levels of the virus in our water supplies due to the inability of our water treatment plants to have filters with sufficiently small micron sizes to filter out the virus.

Some chemicals, such as pesticides, bioaccumulate in organisms, resulting in long-term health effects in higher trophic levels, including humans.

Plastic pollution

We are now living in what anthropologists and geologists call the Anthropocene, a new epoch characterised by the profound impact of human activity on the Earth's geology and ecosystems. Our fingerprints are stamped into everything: the soil, the air, the oceans, even the very rock strata that will one day fossilise our age. And perhaps nothing symbolises this more grimly than plastic.

Plastics, especially single-use waste like bottles, bags, and food packaging, are now so entrenched in our waterways that they could be classified as a new geological layer. They do not decompose; they simply fracture, splintering into tiny shards known as microplastics. These particles are now everywhere, drifting through rivers, swirling in ocean currents, floating in the air, and even falling to the ground in the rain.

Once in the water, microplastics act like chemical sponges, absorbing toxins such as heavy metals, pesticides, and persistent organic pollutants. Plankton consume them, mistaking them for food. Small fish eat the plankton, larger predators eat the fish, and the contamination travels up the food chain until it ends up in the stomachs of marine mammals, seabirds, and finally, on our own plates. We are, quite literally, eating our own waste.

The larger plastic debris is just as lethal. Marine animals become entangled in bags, ropes, and abandoned fishing gear, leading to slow deaths by strangulation, suffocation, or starvation. Turtles choke on plastic bags they mistake for jellyfish. Whales wash ashore with stomachs full of carrier bags and bottle caps, dying with bellies full yet starving for lack of nutrition. Seabirds feed colourful plastic fragments to their chicks, unaware that they are slowly killing them. Coral reefs and seagrass beds, nurseries for marine life, are smothered and broken by drifting nets and discarded sheets of plastic, destroying the very foundations of ocean ecosystems.

And now, entire ecosystems are beginning to unravel.

The floating graveyard — The North Atlantic Garbage Patch

The North Atlantic Garbage Patch, one of five major gyres in the world's oceans, is a swirling soup of plastic debris, trapped by circulating currents. It stretches for thousands of kilometres and is estimated to contain hundreds of thousands of tonnes of waste, much of it fragmented into microplastics invisible to the naked eye. But its impact is impossible to ignore.

And it does not stay at sea. The same oceanic currents that trap plastics eventually push them back to shore. Beaches that were once golden and wild are now streaked with ribbons of waste; in some regions, as much as 5% of beach sand now consists of ground-up plastic fragments.[15]

If the rivers and seas are the veins of the Earth, we are clogging them with plastic blood clots. Every bottle, every shopping bag, every single-use coffee cup becomes part of this planetary suffocation. And while this may seem distant, out there in the deep ocean gyres, it is already in us.

Microplastics are now being detected in human blood, placentas, and breast milk.[16] We are poisoning not just the Earth, but ourselves.

The question is no longer *if* this will affect us, it already has. The only question is whether we are willing to change before the veins of the world stop flowing altogether.

The accompanying image details a North Atlantic Garbage patch—something which no one ever talks about!

---

[15] María L. Jaubet and others, 'Factors Driving the Abundance and Distribution of Microplastics on Sandy Beaches in a Southwest Atlantic Seaside Resort', *Marine Environmental Research*, 171 (2021), 105372.

[16] Antonio Ragusa and others, 'Plasticenta: First Evidence of Microplastics in Human Placenta', *Environment International*, 146 (2021), 106274. Sophie V. L. Leonard and others, 'Microplastics in Human Blood: Polymer Types, Concentrations and Characterisation Using μFTIR', *Environment International*, 188 (2024), 108751.

The map details several Garbage Patches from NASA.

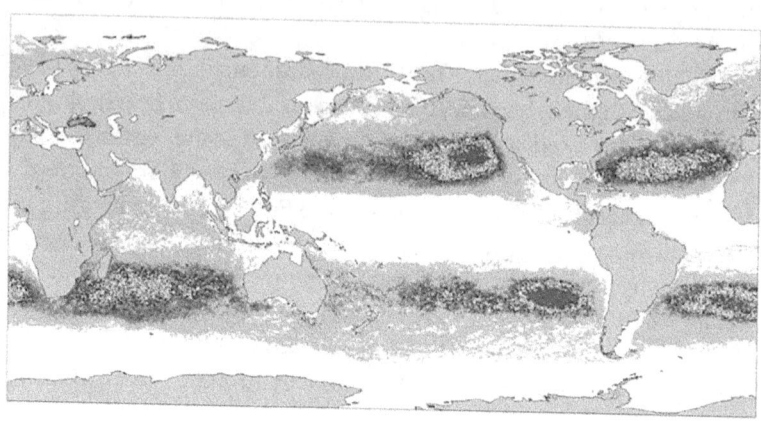

The great recycling delusion

We spend so much time and care sorting our waste, every yoghurt pot rinsed and every piece dutifully placed in the correct bin. It feels righteous, this daily rite of redemption through plastic and paper. We imagine that, by doing so, we are helping to heal the world or at least doing our part.

But while we stand in our kitchens, washing out our pasta sauce jars, the skies are thick with jets. Private flights, many of them taken by celebrities and business elites, have risen by nearly half in just a few short years.[17] In 2023 alone, private aircraft emitted approximately 19.5 million tonnes of greenhouse gases, surpassing the total emissions from all commercial flights departing from Heathrow that year.[18] Some of those jets, owned or chartered by the famous, have clocked up enough air hours to circle the Earth thousands of times. The hypocrisy of the 'green elite' is staggering—they tweet about sustainability while flying solo to collect environmental awards. You would think that this was made up, but nope, 'Leonardo DiCaprio used a private jet to travel

---

[17] International Council on Clean Transportation, 'Private Jet Emissions Increase by 46% between 2019–2023' (2023) <https://theicct.org/pr-air-and-ghg-pollution-from-private-jets-2023-jun25/>.

[18] Damian Carrington, 'The Jet Set: 200 Celebrities' Aircraft Have Flown for Combined Total of 11 Years since 2022' (2023) <https://www.theguardian.com/environment/2023/nov/21/the-jet-set-200-celebrities-aircraft-have-flown-for-combined-total-of-11-years-since-2022>.

from Cannes to New York to accept an environmental award—a move widely criticised as hypocritical given his climate activism.'[19] This was reported by Earth.org, a politically active environmental group.

Meanwhile, we, the faithful recyclers, continue to believe in our cause. But wait, there's more. The evidence shows that a significant portion of this pious effort is wasted. Much of what we carefully separate is quietly shipped overseas, to be dumped, burned, or buried under another sky.[20] Britain's exports of plastic waste have actually increased by 84% in recent years, despite public assurances of domestic recycling. It is a sleight of hand on a national scale, lies disguised as virtue.

The hidden export of guilt

Containers marked 'recyclables' leave British shores, bound for countries with poor or non-existent waste management infrastructure.[21] In places like Malaysia, Turkey and Indonesia, the mountains of imported plastic are left to rot or burn, leaching toxins into rivers and soil. In the United States, too, millions of tonnes of 'recycled' waste are exported each year to the Global South, where up to 85% of it may never be processed at all.[22]

We are told that recycling is our duty, a small personal contribution to the planet's health. Yet it has become a ritual of distraction, a convenient theatre that allows governments and corporations to continue as before. It is not the yoghurt pot that poisons the Earth; it is the engines that roar across the sky, the endless shipping, the hidden export of guilt.

Now, understand that I am not saying we should not recycle; it is a critical thing we should do. However, we cannot recycle our way out

---

[19] Earth.Org, 'Celebrity Culture and Environmental Impact: An Urgent Call for Accountability' (2023) <https://earth.org/celebrity-culture-and-environmental-impact-an-urgent-call-for-accountability/>.
[20] Sandra Reynolds, 'Growth in Plastic Packaging Recycling Being Shipped Abroad, Including Non-OECD Countries' (2024) <https://envirotecmagazine.com/2024/01/24/growth-in-plastic-packaging-recycling-being-shipped-abroad-including-non-oecd-countries/>.
[21] Sandra Laville, 'UK's Plastic Waste May Be Dumped Overseas Instead of Recycled' (2018) <https://bbia.org.uk/uks-plastic-waste-may-dumped-overseas-instead-recycled/>.
[22] Plastic Pollution Coalition, '157,000 Shipping Containers of U.S. Plastic Waste Exported to Countries with Poor Waste Management in 2018' (2019) <https://www.plasticpollutioncoalition.org/blog/2019/3/6/157000-shipping-containers-of-us-plastic-waste-exported-to-countries-with-poor-waste-management-in-2018>.

of this crisis. The problem is not the consumer; it is the system—the endless pursuit of convenience, profit and escape.

Real change will not come from the sorting bin, but from dismantling the illusion that the poor can atone for the rich.

Hydration—The world's greatest sales pitch

We are so soaked in capitalism that we barely notice when we're sold something we already have. Enter Evian, proudly telling us its heroic origin story.

A French nobleman, the Marquis de Lessert, went for a stroll in 1789, got thirsty, and drank from a spring in Évian-les-Bains. "How light! How refreshing!" he said, and promptly made it his daily habit. A local entrepreneur, Monsieur Cachat, then had the bright idea to sell it to his neighbours. People loved it so much they wanted to bathe in it, so spas opened. Bottling plants followed. By the end of the 19th century, this water was being shipped out as the height of health and wellness. By 1978, Danone had shipped the whole farce to America, selling it as 'premium' hydration.

Cue the inspirational music, right?

But here's the punchline: you're paying 100 times the price for something that already flows from your tap.

And just to rub salt in the wound—write 'EVIAN' backwards.

NAÏVE

That's us. That's our collective stupidity wrapped in plastic bottles.

And there's worse news: we leave our bottles on windowsills, stash them in hot cars, and carry them on summer walks, clutching plastic water bottles like talismans of convenience. But under sunlight, those bottles don't just sit there innocently. They react.

When plastic is exposed to UV rays, it begins to break down, undergoing changes at the molecular level. The sun's energy degrades the polymer chains, releasing chemicals into the water they're supposed to contain. Among the worst of these are BPA (bisphenol A), phthalates, and antimony, chemicals linked to hormonal disruption, reproductive issues, developmental delays in children, and increased cancer risk.[23]

---

[23] W. Shotyk and M. Krachler, 'Contamination of Bottled Waters with Antimony Leaching from Polyethylene Terephthalate (PET) Increases Upon Storage', *Environmental Science and Technology Journal*, 41(5) (2007), 1560–63.

What's terrifying is that this isn't dramatic. It's subtle. It's slow. You won't taste it, or smell it. You'll just drink what seems like clean water, and slowly absorb a toxic legacy.

Over time, regular exposure to leached plastic chemicals contributes to the growing epidemic of endocrine disorders, fertility issues, and immune system dysfunction. The human body, like the rivers and oceans, was never meant to be a container for industrial waste.

And let's not forget, we've known this since 2007. Scientists warned us three decades ago that sunlight causes plastic bottles to leach chemicals into our water. Yet, we're still given the same hollow reassurances: *plastic is safe. It's recyclable. The problem is you, not the packaging.*

But let's stop pretending. The truth is as clear as the water in that bottle: every time you drink from sun-warmed plastic, you're not just quenching your thirst, you're swallowing the price of convenience, one toxic sip at a time.

The way forward...

Gathering in the wild is not a luxury for us; it is essential. It feeds us in ways no city, no machine, no screen ever can. To stand barefoot in the grass, to feel the hum of a forest, to sit with the slow patience of a stone or the whisper of a stream—this is medicine. And it isn't just poetic nonsense; science has finally caught up.

Post-COVID research on the psychological and physiological effects of Nature now echoes what Pagans have known for centuries. Japanese studies on *forest bathing* list the benefits in clinical detail: improved heart health, lowered inflammation, boosted immunity, reduced anxiety, eased depression, lifted moods. The whole list is long enough to make any doctor weep with joy.[24]

And yes, my first thought was, "No shit, Sherlock." We didn't need a white coat and a research grant to tell us that being in the forest is good for the soul. We've been saying it, and living it, forever.

But here's the hard truth: loving the land isn't enough. It's not enough to just enjoy nature, to soak up its magic, to dance under the trees, and then walk away leaving it worse than we found it. If we are

---

[24] Ye Wen and others, 'Medical Empirical Research on Forest Bathing (*Shinrin-Yoku*): A Systematic Review', *Environmental Health and Preventive Medicine*, 24(1) (2019), 70.

to call ourselves Pagans, animists, or just human beings with a shred of integrity, we need to step up.

The hard questions we must ask

Every ritual, every gathering, every camp, every time we step foot on sacred land, we should be asking ourselves:

- What impact are we having on this place?
- What are we doing to the other beings who live here, our green friends, our non-human neighbours, the ecosystem itself?
- And most importantly, how can we put it right?

This isn't optional. It's the very heart of what it means to be an animist, Pagan, witch, druid, whatever you call yourself. We are in a symbiotic relationship with the land. If we take without care, if we leave scars, if we treat sacred sites like festivals rather than temples, the land will not continue to give.

The first steps

Make restoration part of your magic. Build it into your ritual. That means:

- Organised clean-ups. Pick up after yourselves and others. Sacred sites should be left better than you found them.
- Purchase glass drinking bottles and educate others about the potential damage caused by using plastic bottles.
- Respect for fire. Research fire-lighting, use it sparingly, and know how to erase your traces.
- Rethink offerings. Join groups like the Cleaner Clootie Campaign—remove synthetic clooties and plastic rubbish strangling sacred wells and trees. Offerings should be biodegradable, meaningful, and safe for wildlife.
- Raise your voice. Challenge damaging behaviour. Contact organisations like the Pagan Federation or the Order of Bards and Druids, and urge them to guide their members towards truly respectful practice.

And do not stay silent. Too often, we whisper, too polite to confront the damage done by our own community.

Ecological magic is not only about personal growth or quiet contemplation; it is part of a greater task. We are living in a time when the green agenda is faltering. Governments and corporations claim to promote sustainability while destroying forests and polluting rivers. Green industries—those fragile, necessary attempts at balance—are too often sacrificed for short-term profit.

If we are witches, if we are workers of the land, then we cannot be neutral in this. Our rituals must be mirrored by our physical efforts. That means growing food in ways that honour the soil, supporting rewilding and green energy, and fighting against practices that strip the earth bare. It means choosing where we spend our money, lending our voices to environmental movements, and teaching others that the Earth is not a resource, it is a living being.

Ritual lies at the very heart of Paganism, not as performance, nor as empty repetition, but as a living act of alignment. It is the heartbeat of the path, pulsing through our bodies and through the land. Whether we are marking the shifting dance of sun and shadow across the Wheel of the Year, or working magic beneath the silver pull of the Full Moon, ritual is how we come into right relationship, with the cosmos, with each other, and with the more-than-human world.

Cheal and Leverick have spoken of ritual as a transformative act, and rightly so. They write, *At the centre of every major Neo-Pagan ritual there is some process of transformation, consisting of a change from one state to another.*[25] In the rites of the Sabbats, this transformation is mirrored in the cyclical rebirth of the Earth herself, seasons turning, seeds quickening, leaves falling, all affirming the sacred rhythm of life, death, and return. These rituals are not merely calendars in costume; they are embodied affirmations of vitality, and I honestly believe they feed the land where the rituals are held. They offer, in every solstice fire and Beltane dance, a quiet but resolute renewal of self and place.

Yet I would argue that perhaps the deeper alchemy, the one that cracks the soul wide open, emerges more often in the Full Moon rites. Here, I have witnessed authentic personal gnosis unfold: moments of emotional catharsis that heal old wounds, encounters with spirit and archetype that leave a lasting imprint on the psyche. These are

---

[25] David Cheal and Jane Leverick, 'Working Magic in Neo-Paganism', *Journal of Ritual Studies*, 13 (1999), 7–19.

not abstract transformations, but visceral, lived shifts in identity, understanding, and purpose. They are, to borrow the term, *soul-making*.

Charmain Sonnex likens ritual to *eudaimonia*, not simple happiness, but a life imbued with purpose and honour. She describes ritual as a liminal flow that draws us back into a life infused with soul—a life that feels worthwhile, carried with dignity and presence. And she links this to Aristotle's concept of *eudaimonia*, as explored in the *Nicomachean Ethics*. This flourishing comes when one lives in alignment with one's inner *daimon*, one's true and authentic self. This flourishing, she suggests, consists of six elements: personal growth, self-acceptance, positive relations with others, autonomy, environmental empathy, and purpose.[26]

I have observed this through personal experience on the Pagan path. Personal growth and self-acceptance are common outcomes of sustained practice. Relationships deepen, not only with fellow humans but with the spirits of land and place, with the ancestors, with stone and stream. Paganism encourages autonomy and the rejection of centralised dogma; there is no final book, no rigid creed, only the living tapestry of experience, story, and spirit. The path is one of deep, rooted purpose.

There is also a long and honourable tradition of Pagan ritual as a form of resistance. We are not simply dreamers dancing in woods; we are spell-workers standing against destruction, weaving threads of power in the name of justice and renewal.

One such example took place in 1982, when a group of seventy Pagans gathered to perform a rite against nuclear escalation. *Pagans Against Nukes* was not just a symbolic gesture. It was a magical act, rooted in political urgency. Ferrero and White describe it as a simple ritual, using chants such as Starhawk's *"We can rise with the fire of freedom."* The simplicity did not reduce its power—it concentrated it.

Another potent rite of our time is the *Council of All Beings*, born of Deep Ecology and first brought into the world by Joanna Macy and John Seed. This ritual is not about invoking deities or casting circles in the traditional sense; it is about stepping aside from human identity and allowing the other-than-human voices to speak through us.

In this council, participants become trees, rivers, fungi, and whales. They speak with the voice of moss, or crow, or the poisoned Earth herself. Grief is given room to rise. Insight is drawn from the bones of the world.

---

[26] Charmaine Sonnex and others, 'Flow, Liminality, and Eudaimonia: Pagan Ritual Practice as a Gateway to a Life with Meaning', *Journal of Humanistic Psychology*, 62 (2022), 233–56.

It is not always comfortable, but it is always real. The five stages of the rite—setting intention, mourning, ecological remembering, council dialogue, and reintegration—create a container in which the boundary between human and nature dissolves.

It is not performance. It is prayer with teeth.

Those who take part often leave changed. They speak of a deepened ecological ethic, of hearing the voice of the Earth in a new way. Grief is no longer a private burden; it becomes a communal offering. This, too, is what ritual can do: it reminds us we are not alone.

I participated in this ritual at the Mercian Gathering Camp in the UK, and it had a profound impact on both my own practice and the practices of other pagans who attended. Many had an emotional reaction to the ritual, and from talking to people since participating, it has had a profound effect in encouraging pagans to engage in physical protest and resistance.

These examples, ecstatic, political, and ecological, demonstrate the central role of ritual in Pagan life. It teaches, it heals, it protests, it remakes the self in the image of a world worth saving. It would be easy to paint Paganism, then, as the 'great green hope', a spiritual ecology in action.

And in many ways, it is.

But hope alone is not enough. There are still shadows to confront: performative greenwashing, consumable spirituality, and rituals that celebrate the Earth while still trampling her underfoot. The work is not done simply by casting the circle; it is done by walking the talk, in mud and in magic alike.

I include different types of rituals later in the book; some are developmental, and some are ecoactive magical rituals. It is my feeling that ritual is not an escape from the world. It is a doorway back into it.

PART TWO

BLACK PATHS AND GREEN CATHEDRALS

CHAPTER FOUR

Working with the land spirits

Making a pact with the land

This is probably one of the most critical magical acts you can perform. The land on which we work magically is the source of our being. All other relationships are secondary to it, for it is here that we live, breathe, and have our being.

Contacting the *genius loci* of your area is not a complicated ritual. I will now provide an example of my group's practice, and you can adjust it as needed. We begin with a simple question: How do I get to know my land?

To get to know the land around you, you must physically and mentally venture out. A great way to begin is by answering the following questions.

- When does the first leaf appear on the trees?
- What type of tree blossoms first?
- Which is the first animal you see when winter is over?
- What geological process formed the land where you live?
- What was the total precipitation in your area last year?

- Name three or more birds that are common to your area, and which live there all year-round; which ones migrate?
- Name three or more mammals which are common in your area.
- Which wind blows the most winter storms into your area?
- What are the primary sources of pollution in your area?
- Where is your closest natural source of water?
- What type of soil do you have in the area?
- When was the last Full Moon?
- If you stand outside at night, what 'natural' sounds do you hear? Do this in the spring, summer, autumn, and winter.

The *genius loci*, the spirit of place, is not just some abstract idea; it is a living, breathing presence, and one you should absolutely make the effort to get to know. It weaves itself through everything in your immediate world: your house, your garden, the hedges, the soil, even the scruffy weeds pushing through the pavement cracks.

The longer you live in a place, the stronger this relationship can become, and the more you'll learn from it. Over time, this spirit settles into you as much as you settle into it. It's what makes a place feel truly like Home, not just somewhere you happen to live, but somewhere that recognises you back.

So, with that in mind, I'll share a short but powerful ritual for making contact with the Genius Loci, because if you're going to share space with it, you might as well introduce yourself properly.

A short ritual to the Genius Loci for contacting the Spirit(s) of the Place and Plant(s)

Time: On a New Moon or when the Moon is waxing. If possible, when Mercury is in a good aspect to the Moon or when the Moon is waxing in Gemini.

Place: Go out into the woods or your place of choice—this can be anywhere that is special to you. Safety first. Ensure you are not disturbed; people tend to look askance at someone in full robes in the middle of nowhere, so be subtle; jeans and a jumper are a better option.

Method: Visualise yourself within the enclave of your Aura, see it as a shield against any harm, and know that you are safe within. (NB we do not use a circle in this circumstance—we want the spirits to come, and by making a circle, you place a barrier to those spirits.) However, if you are uncomfortable with this approach, please consider using a circle instead, but note that your contact may be reduced.

Invocation: As you see fit; use either your own deities or ones that you have researched. DragonOak is dedicated to Hekate and Hermes, so in this ritual we begin by calling upon those gods—use a fervent and active voice, or simply chant the following:

Invocation to Hekate of the Green

> *"Mother of the form around me*
> *Life and death in one deep breath*
> *Universal hearts compassion*
> *Mother of both birth and death.*
> *Cup and womb of power indwelling*
> *Give to me the kingdoms key*
> > *Shadow to the lights great power*
> > *Mother of the world to be.*
> > *Scythed one standing on the crossroad*
> > *In between the dark and light,*
> > *Give to me an ancient power*
>
> *Grant to me the ancient sight.*
> *Hekate of the Green, Physis, Chthonia,*
> *I ask you to be a bridge between myself and the spirit who dwell in this place*
> *Be a doorway through which I may pass into understanding*
> *Make communication easy and let us form a bond."*

Invocation to Hermes of the wood — asking for aid in communication

> *"Serpent coiled around the tree branch*
> *Seed of light within my soul*
> *Shine upon these rites eternal*
> *Grant me sight of wisdom's goal*
> *Horned one standing in the greenwood*
> *Green boughs hanging from the trees*
> *Show me how to talk to others*
> *Send my message to the green."*

The call to your spirit

> *"Spirits, guardians of this place, in this time and in this place*
> *I call to you*
> *Come for I would know you better*
> *Come and let me learn from you*
> *Spirit of the trees and plants*
> *I ask you to bless this work and give me your aid.*
> *I come with an open heart and with good intent.*
> *I offer service and companionship."*

The Offering: Now is the time to give something back. This could be incense, tea, a splash of spirits, or—if you've done your homework—something the local spirits actually want. Research is good. And if you're in a wooded area? Urine is an excellent offering, packed with lovely nitrates that plants slurp up like a fine wine. (Just remember: pick your spot wisely. Offending the spirits is bad; offending the dog walkers or getting arrested for public indecency is arguably worse. Be discreet unless you fancy explaining to the police that you were 'feeding the trees'.)

Focus and internalisation: Open your heart centre and put the call out with your emotional senses. When the *genius loci* or a plant spirit comes, you will feel it. However, don't expect to understand; just let your intuition work and remember the feeling. For the next time you visit the area, you should feel welcome. Indeed, the more you go and perform rituals, the more you help the area grow, and the stronger it will become. Your experience will also become even more welcoming.

Now—Putting your money where your mouth is...

You will notice from the above ritual that a major part of it involves being of service; this is essential. You must work with the land, and then, after telling the land spirits that you will be of service, you must live up to that end of any bargain that is struck. So, activities such as litter picking and cleaning the area, ensuring that other pagans' practices are properly cleaned up, and maintaining the land in excellent condition, all contribute to a well-maintained area.

Planting trees and engaging in a bit of Ninja gardening is also highly recommended; however, ensure that you are growing plants that are in line with the environment you have been working with. For example, avoid planting palm trees in Treorchy.

You can also work with local ecological groups by volunteering to upkeep a place of scientific or natural beauty. This is one of the best ways to truly 'pay service' to the land you work with. In Wales, where we often hold our sabbat rituals, we make it a point to give back, picking up rubbish, tending to the gardens, and ensuring the land is treated with respect. It isn't just about keeping things tidy; it's about building a deeper, more magical relationship with that place.

We talk to the land, we ask it what it needs to thrive, and we listen to the *genius loci* when it answers. This is a two-way relationship, after all. The land gives to us; we should be giving back.

We try to work in that area as often as we can. It's a commitment, yes, but a sacred one, an intention to place and space. Over time, that intention becomes something tangible. The land begins to know you, to trust you, and in turn, your magic there deepens in ways you can't get from a once-a-year visit.

Creating a more personalised relationship with Nature

Use what grows beneath your feet

One of the greatest acts of ecological paganism, and perhaps one of the most potent acts of devotion to the land, is to stop reaching across oceans for magic that already grows beneath your feet. Why are we importing plants sacred to other peoples, tearing them from their native lands, when the spirits of our own soil are waiting for us?

Take mugwort, for example. We buy white sage by the bundle, stripped from Indigenous lands where it is overharvested and sold as a 'witchy accessory', while mugwort grows wild and abundantly all across the UK. She makes a beautiful smudge stick, her smoke soft and silvery, wrapping around you like a whispered secret. She opens the gates of dreaming as readily as white sage ever could—perhaps more so, because she knows *you*. She knows the damp, rich soil you walk on. She belongs to this land, and if you work with her, she will teach you how to belong too.

The same goes for sacred plant allies used for journeying. People travel across continents chasing ayahuasca, treating it as some mystical tourist experience, when the spirits of our own land already hold the keys to those same doors. Psilocybin mushrooms—the little brown caps that rise from sheep-grazed fields—have been used in European magic for millennia. You don't need to fly to the Amazon. You need to walk the hills where the sheep graze, tread gently among the grass, and learn the spirit of the mushroom as an honoured teacher, not a party trick. Take it in ceremony, with reverence, and you will walk the paths of the gods just as surely as any South American shaman.

And why ship frankincense across deserts when your local pine trees bleed resin that burns with a soft, clean smoke? Pine resin is abundant, easy to collect without harm, and holds a bright, clearing energy of its own. It carries the wildness of the forest, the sharp green breath of the northern lands.

Using local plants is not just a matter of convenience, it is about *relationship*. It is honouring the land that feeds you, shelters you, and carries your footprints every day. These plants are not strangers; they know the wind that tangles your hair, the rain that soaks your skin. They are part of your story.

If you want your magic to matter, if you want it to be real—stop buying spirituality in plastic packets. Go outside. Learn the plants that grow under your own sky. Work with them as kin, not commodities. That is where the real magic begins.

If you're a witch or pagan who uses magical tools, then why not actually make them yourself? There's nothing more ecological, or more satisfying, than crafting your own equipment. When you shape a wand, carve a staff, or forge a blade with your own hands, you're not just making an object; you're weaving yourself into it. The tool becomes an extension of you, and you become an extension of it.

And here's the best bit: it gets you outside. It pushes you into the wild to meet the beings who live there, to ask permission, to build relationships with the spirits of wood, stone, or metal. That's real magical work, not just clicking 'buy now'.

Because let's be honest, walking into a shop full of shiny witchy trinkets is tempting, but giving in to that temptation is a weakening of will. If you can't resist a pretty bauble, how can you claim to have mastery of your own magical direction? True craft takes effort. It takes time. And the land, and the spirits you work with, will respect you far more for it.

In my coven, DragonOak, we craft our robes, cingulum, wand, pentacle, cup, and blade.

At this point, I want to concentrate on the magical wand rather than on our other tools, because ecologically, you are more likely to buy a knife or cup. Still, even here, I suggest you buy one from a second-hand shop and cleanse and purify it to make a more personalised and magical object. The *wand* is primarily an instrument of nature, and rather than spending money on something that is created by someone else or may have simply been taken without thought for the tree, you should endeavour to find and make your own wand. This is achieved by first connecting directly with the tree.

Establishing a relationship with a tree

Further on in this book, you will find details on how to form a relationship with a tree, as well as a list of the trees I have specifically worked with. These examples will provide practitioners with a deeper understanding of their use. Yet, before I get to this, I want to talk to you briefly about the word 'NO'.

In our anthropocentric worldview, we very rarely hear or internalise the word 'no'.

Even more importantly, when we *do* hear 'no', we often ignore it. Why? Because we *want* the thing. Because we're human, and we think we're entitled to it. That attitude—"I am human, therefore I may take"—is poison to any real magical relationship.

When you take a wand, or any part of a plant, you are beginning a magical partnership, not committing a casual act of harvesting. That relationship takes time. It takes patience. Simply hacking off a branch is not wand-making; it's an act of violence. A tree is not a shop, and magic is not a form of consumer culture. If you approach a tree as if you're just 'collecting supplies', don't expect your wand to have any spirit, any power, or any bond with you.

The right way is slower, deeper, and infinitely more rewarding. Sit with the tree. Listen. Feel. Form a real relationship with the spirit of that being. Sometimes this takes weeks. With elders like the yew, it can take months. But I promise you, the effort you invest will pay magical dividends that a rushed, grab-and-go wand never will.

How to ask

When you're ready to ask for permission, use your voice. Introduce yourself, as you would to any other sentient being:

> "Hello, my name is Sian Sibley. I'm a magical practitioner, and I want to deepen my relationship with this land and its spirits."

Then ask, clearly and respectfully:

> "May I take a part of you to join me in magical work?"

Now, listen. Really listen. Plants and trees speak in many ways, but when the answer is **NO**, you'll almost always feel it in your body. Many practitioners experience a sudden tightening or pressure in the chest, or an energetic 'push' in the heart space. It's a very physical, unmistakable feeling, like the plant is quite literally pushing you away. When you feel that, you *know*. And when you know, you walk away.

Respect that answer. Thank the tree, leave it in peace, and move on to ask elsewhere.

Why this matters

This practice does three important things:

1. It builds a real connection to your land. You will learn the trees and plants of your area intimately, not just as 'materials' but as allies and neighbours.
2. It sharpens your intuition. Over time, you'll learn how different plants speak to you, whether they give you a firm 'no' in the chest, a quiet 'yes' in your gut, or something more complex.
3. It enhances both your magical and physical well-being. Sitting quietly among trees, listening, waiting, being—this is magic in itself.

Only after this relationship has been formed, only after you have been given direct permission, should you cut your wand. And when you do, cut it with respect, gratitude, and a promise to honour that tree every time you use it.

Preparation

You will need to prepare and take the following items with you:

1. a SHARP knife or hand saw to cut the branch;
2. a Gift for the spirit—I tend to offer blood or urine as an exchange';
3. a clean piece of cloth in which to wrap your wands (cotton, wool, or silk is preferable);
4. some beeswax to seal the end of the cut branch to stop any infection from occurring.

Method

1. Now, reach out and touch the tree. Allow your energies to entwine directly with the tree ... draw in, and then
2. experience a direct union with the tree.
3. Cut one branch to the approximate length. from your middle finger to your elbow; this branch should be as thick as your thumb.
4. Wrap the branch in the cloth you have brought.
5. Thank the tree for its loving kindness and its generosity.

Making the wand

When you arrive home, you can either carefully strip the bark from the wand or leave it in its natural form. Whether you decide to strip or simply polish the bark, you can communicate with the spirit that remains embodied in the wand. You can sing to it, but most importantly, you must give it your full attention and thank it for working with you.

Finishing and activating your wand

At the New Moon, when you are ready to finish your wand, gather the following materials:

1. a beeswax candle and matches;
2. a clean cloth;
3. a sterile needle;
4. a cotton ball;
5. a sharp, pointed object to bore a hole in the end of the wand;
6. any other ritual equipment that you want to use if you decide to consecrate to purpose.

Now—light the candle

Hold the wand in your hand (depending on which hand you use naturally—I use my right hand), ground and centre.

Connect with the tree spirit and allow your energies to mingle.

When the energy is flowing nicely, you may begin shaping the wand.

The narrow end is the projective end. It may be carved in either a phallic or cone-shaped design. You may also simply round it off.

The wider end is the receiver end. To this end, drill a small hole using a sharp, pointed object.

Using the sterilised needle or lancet, prick your finger and drop your blood into the hole you have made, then seal the hole with melted wax from your candle.

This part of the ritual symbolises the mixing of your spirit with that of the tree's spirit. It is essential as it seals the wand to your use (as you originally did with the tree). This is why a shop-bought wand will never be as powerful an ally as one made in this magical way.

Afterwards, you may dedicate the wand to a particular use or deity, or you may also just keep the wand for more general magical use. Wands can also be given a name (this again forms and strengthens a very personalised interrelationship between you and the tree—hence, when used it becomes an interconnective force in its own right).

Incense, resins, and herbs

You can create wonderful incense and resins from local plants; you don't need to rely on expensive, shop-bought blends. Pine resin is one of the easiest and most magical to gather, and it's commonly found in older, more natural woodlands. But, as with wand-making, you must treat this as a relationship, not a raid. Always ask the plant for permission before taking anything. For more on forming these relationships, see my earlier book, *Unveiling the Green*. You'll find that many plants are more than willing to work with us and will happily teach us, if we only listen.

A quick practical note on pine resin: only take it when it's solid on the trunk. If you try to scoop up runny resin, you will regret it, as it will coat your bag, your clothes, and anything else it comes into contact with. Trust me, it's a sticky nightmare.

Best local trees for resin

1. Scots Pine (*Pinus sylvestris*)
   Resin: Abundant and easy to collect—look for hardened resin on older trunks or around healed wounds.
   Magical use: Cleansing, purification, protection, and grounding. Excellent for clearing ritual spaces.
2. Norway spruce (*Picea abies*) (also Sitka spruce in some regions)
   Resin: Produces fragrant resin, often dripping from natural fissures in the bark.
   Magical use: Healing and protective workings, good for house blessing and warding.
3. Douglas fir (*Pseudotsuga menziesii*)
   Resin: Less common, but when found it burns with a sweet, uplifting scent.
   Magical use: Joy, renewal, calling in good spirits; often used for ritual invocations.

Take some time to investigate what grows in your area—your local land is already offering you magical allies if you just look. For example:

Birch and alder bark make a wonderfully subtle incense with a soft, earthy aroma. Take care to ensure the room is well ventilated when using birch bark.

Cedar bark, which is often shed naturally and can be found on the ground around cedar trees, makes a beautifully cleansing incense. It's perfect for purifying and grounding work, and, best of all, you can collect it without harming the tree.

And if local resins are scarce, don't despair, your kitchen spice rack is already a magical apothecary. Cinnamon, lavender, bay and rosemary smell fantastic and burn beautifully. Wild herb gathering and resin collection are also excellent ways to get outside, connect with the plants, and build genuine relationships.

Mugwort smudge sticks—A better, wilder alternative to white sage

As discussed earlier, white sage has been overharvested to the point of ecological damage, and it doesn't grow natively in the UK. But mugwort (*Artemisia vulgaris*), one of our oldest magical allies, is abundant, powerful, and ideally suited for cleansing and spirit work.

Mugwort is an ancient herb associated with witches, and used for protection, divination, and opening the gates of dreams and visions. It's deeply tied to British and European folklore, making it an ideal local alternative for ritual smoke cleansing. And unlike white sage, gathering responsibly actually strengthens your relationship with your own land.

When and how to harvest mugwort

Ask first—As always, ask permission. Mugwort is generous but not a doormat; take a moment to connect, introduce yourself, and feel for that 'yes' in your gut. If you feel a heavy 'push' in your chest or heart space, it's a 'no'. Thank the plant and move on.

The best time to harvest is late summer, just before flowering, which is ideal. The leaves are at their strongest and richest in oils. However, you can harvest smaller amounts any time it's abundant. I actually like making mine with the flowers as well, as I like to use it in smoking, for dreamwork.

How much to take? Never strip a plant bare. A good rule: never take more than one-third from any single plant, and only harvest from healthy, thriving stands.

Cut, don't tear. Use a sharp knife or pruning shears; tearing bruises the plant and wastes energy.

Thank the plant. Leave a small offering if appropriate (a whispered prayer, a strand of hair, or a drop of water or urine at the roots).

Making your mugwort smudge stick

You'll need:

- fresh mugwort stems (long and pliable: 15–20 cm is good);
- natural cotton or hemp string (undyed);
- scissors.

Step 1: Bundle the stems
Gather 6–10 stems into a neat bunch, aligning the bases as you go. For a thicker smudge stick, you can add more, but keep it tight enough to dry evenly.

Step 2: Wrap the bundle
Start at the base and wrap your string firmly, working your way up to the tip, then back down again in a criss-cross pattern.

Tie securely at the base.

Step 3: Drying
Hang the bundle upside down in a warm, dry, well-ventilated place for 2–3 weeks.

You'll know it's ready when the stems snap rather than bend.

Using mugwort smudge sticks

Light the tip gently and let it smoulder. Mugwort produces a soft, earthy smoke with a faintly bitter edge.

It's excellent for cleansing spaces, tools, or yourself before ritual.

Mugwort smoke is perfect for divination, dreamwork, and trance states, but go easy. Its spirit is potent, and too much can be overwhelming.

Alternatives to crystals in magical practice

Working with local materials isn't just about being ecological; it's about deepening your relationship with the land under your feet. As we previously discussed, crystals often come from far away, and are mined in ways that strip both land and spirit. In the UK, we're surrounded by natural stone allies. Using them in magical practice doesn't just connect you to the Earth in a general sense; it roots you to *your* land, to the land spirits (*genius loci*), and to the quiet magic that has always been here.

Local stone allies for magical practice

Flint

<u>Where to find</u>: Common in chalk downs, fields, and coastal areas (especially in the south and east of England).
<u>Magical use</u>: Ancient protective charm; sparks fire and energy; used historically for warding off evil spirits and for ancestor work due to its deep prehistoric associations.

Slate

<u>Where to find</u>: Wales, Cornwall, and Northern England.
<u>Magical use</u>: Grounding and protection; good for inscribing sigils or runes, its flat surface makes an excellent natural altar pentacle.

Jet (fossilised wood)

<u>Where to find</u>: Particularly along the Yorkshire coast (Whitby).
<u>Magical use</u>: Powerful for protection, grief work, and connecting with the ancestors; historically used in mourning jewellery for its strong underworld associations.

Granite

<u>Where to find</u>: Common in Scotland, Wales, and parts of Cornwall.
<u>Magical use</u>: Stability, strength, endurance. A solid foundation for any ritual that requires resilience or long-term focus.

Pebbles from local rivers or streams

<u>Magical use</u>: Excellent for emotional healing, cleansing, and divination. Each carries the memory of its journey through water. Ensure you take care not to affect the wildlife here.

Sea glass or tumbled coastal pebbles

<u>Where to find</u>: Beaches and tidal pools.
<u>Magical use</u>: Transformation, emotional clarity, and connection to the sea's spirit; perfect for water magic.

Quartz

<u>Where to find</u>: Found locally in the UK, scattered in streams, fields, and coastal cliffs.
<u>Magical use</u>: Amplification of energy, cleansing, and divination; every bit as potent as imported quartz, just far more connected to your land.

Hagstones

Hagstones, which are naturally holed stones found on beaches, riverbeds, and chalk streams, have long been steeped in magic and myth. Folklore tells us they were made by fairies or witches, the holes bored through by their spells, or burned by the touch of their power. In some coastal traditions, it was said the sea itself pierced the stones with its endless turning, marking them as gifts from the Otherworld.

Hagstones have always been seen as powerful charms of protection, hung in homes, barns, or on boats to ward off evil spirits and witchcraft. Others were used for spirit sight—peering through the hole was said to let you glimpse faeries, the dead, or hidden spirits walking in the land. Whether shaped by tide, time, or magic, hagstones feel undeniably otherworldly; they are the kind of gift you don't just find—you're allowed to find.

Protection: Hagstones have long been used against witches, evil spirits, and other ne'er-do-wells. They were believed to ward off negative energies and serve as a shield against spells and magic.[27]

Divination: Some traditions hold that hagstones can be used for divination. By looking through the hole in the stone, one could gain insight and guidance.[28]

---

[27] Owen Davies and Ceri Houlbrook, 'Seeking Protection: Objects of Power', in *Building Magic: Ritual and Re-Enchantment in Post-Medieval Structures*, ed. Owen Davies and Ceri Houlbrook (Cham: Springer International Publishing, 2021), pp. 95–119.
[28] Moses Gaster, Book Reviews, *Folklore*, 42 (1931), 485–87.

Healing: It is believed that hagstones can aid in the healing of snake bites. Additionally, they were thought to enable the wearer to see through the disguise of a witch or fairy. There is also some conflation here with the mythology surrounding the snake stones. Snake stones were particularly valued in Welsh culture for their healing power.

Two types were used. One was the result of mass snake mating, where a stone could be found made of the solidified mucus produced by the snakes. Although I prefer Mhara Starling's version of this myth, which describes wise snakes attending a snake conference where they create the stone with their spit through animated conversations.

A second type of snake stone was an ammonite fossil, which, when viewed in the accompanying image, is immediately recognisable as such.

Image: thanks to Springer International Publishing.

Folklore and medical texts from the medieval era indicate that these stones were used by skilled medical practitioners to aid in the treatment of various diseases.

Good luck: The theory goes that only good things can pass through the hole in a hagstone. Therefore, it is believed that hagstones bring good fortune and good wishes to the bearer, while bad luck and evil thoughts are too big to pass through the hole and therefore become stuck in the middle.

Ancestral wisdom: In modern times, hagstones are associated with ancestral wisdom. They are believed to carry the energy of insight and guidance from ancestors. I can find no reference to this in folklore, so I assume it is perhaps more of a modern practice, but as I say, if it helps one to connect, then the method behind it is sound.

Weather witches: One of the lesser-known uses of the hagstone is to aid sailors at sea. Weather witches controlled the wind by using a hagstone attached to a length of sailors' rope, and where the witch would then stand at the top of the cliff overlooking the sea in the direction that the sailors or fishermen were taking their boats. The witch would then swing the hagstone around in a circular motion—if the witch wanted to increase the winds, the stone was swung slowly and in a clockwise direction with increasing speed; and to reduce the winds, the witch swung the stone from a fast position to a slow one in an anticlockwise direction.[29]

At DragonOak, we have used this magical technique in experimental work, and it does seem to have an effect; however, I suggest that a strong rope is needed, as when one generates the force by swinging even a tiny hagstone, it will snap the string and send your hag stone flying off into the unknown!

Finding hagstones: They can often be found along ocean beaches, rivers, and streams, having formed over many years through erosion by water and by a small arthropod called a piddock, which works on the stone to create the holes.

Meditation on the hagstone: I am lucky enough to live in an area about 15 minutes from the coast. I have several hagstones that I work with. I often discuss the persistence of life with them.

Life is a person. It has will, desire; it wants to be everywhere. It pushes out through everything it can, shaping itself into many forms. Some are vast, like the planet itself; some are tiny, like me and like the hagstone I hold.

The hagstone whispers its own questions: What if reality is alive? What if the most ordinary life is planetary? Maybe we're not separate at all—maybe we are simply expressions of the Earth's own personality, its hunger to keep on living. And what of the stars, the other planets— are they alive too? Are we, as Earth-born beings, just the thoughts or

---

[29] Annie Thwaite, 'A History of Amulets in Ten Objects', *Science Museum Group Journal*, 11 (2023).

dreams of a living Sun and a living Earth, moving through us to feel themselves alive?

I mean, just look at this ...![30]

Take just two minutes to meditate on the image.

This isn't a picture of stars; it's one of galaxies, great spiralling cities of light. Within each galaxy are stars, and around those stars, planets turn. On each planet, life expresses itself, though not as it does here. Its forms will be different, strange to us, but life *will* be there, because life is stubborn. It insists. It wants to be. Perhaps, scattered across billions of systems, there are worlds rich with their own green glory, singing their own songs of being alive.

Looking through the hole of a hagstone is like peering into that endless universe, slipping through space and time. I believe life cannot exist without other life to live on, and so everything, in its way, is alive. Even rocks have their own slow, patient life. Where we might live

---

[30] Photograph from the Hubble Deep Field project, NASA.

for ninety years, a rock breathes out its existence over billions of years, shaping itself grain by grain, a quiet gift from the planet to itself.

This is why I urge you to work with the stones of your own land instead of imported crystals. But, here's the catch, if we all rush out to beaches and rivers, gathering stones in great handfuls, we'll only repeat the same harm we've already done to the land. Take only what you need, and when your magical work is finished, return the stone to where it was born.

I do this often, and it changes how I work. Giving the stones back is a way of honouring them; it keeps the relationship alive. When I'm not asking for their help, they are where they belong—back home, breathing with the land that shaped them.

The Black Path

The importance of soil in ecological magic

Every inch of me shall perish. Except one. An inch ... it is small and it is fragile ... Alan Moore, V for Vendetta

This line from the amazing graphic novel *V for Vendetta* (if you haven't read it or seen the film, *Shame on You!*), sums up our relationship with soil, though I don't suppose Moore was thinking about that when he wrote the line.

We live on a planet on which life exists. This is the first miracle in the world. We live because of the few inches of soil, which is the basis of all life on the planet. I am not exaggerating here! Without the soil, there would be no food, trees, animals or us. It is truly where we live and have our being.

From his glorious book *Soil*, Matthew Evans notes that:

> *We can slow down climate change, grow an abundance of better-tasting, nutrient-dense food, and gradually alter the genetic makeup of our bodies for the better and heal the world. The answer is right in front of us, all around and underneath us. All we need to do is stop treating Soil like dirt.*[31]

---

[31] Matthew Evans, *Soil: The Incredible Story of What Keeps the Earth, and Us, Healthy* (London: Murdoch Books, 2022).

The living soil: From sacred earth to silent dust

The soil is not dirt. It is not 'just' the ground beneath our feet, some inert substrate to be ploughed, poisoned, and wrung dry. The soil is alive. It breathes, it remembers, it is a vast community of beings. Fungi, bacteria, invertebrates, plant roots, and microfauna are all weaving the threads of life together in ways we are only beginning to understand.

Once, we knew this. Farmers worked with the soil, not against it. They understood, through a long and patient relationship, that fertility came from keeping the land alive. Manure from cattle, sheep, and horses was spread to replenish the soil's nutrients. Ash from hearth fires was worked into the earth to restore minerals. Green manures—clovers, vetches, and other nitrogen-fixing plants—were sown and then dug back in to feed the next crop. Fields were rotated—grain one year, legumes the next, roots the year after—so the soil could rest and renew. Hedgerows were left standing, providing habitat for pollinators and predators of pests. Mixed planting, or polyculture, mimicked the diversity of Nature, keeping pests in check without the need for chemical warfare.

The land was not always treated kindly—overgrazing and deforestation have long histories—but on the whole, people knew that the soil could not be pushed beyond its limits without consequence. It was understood as a living thing, and living things can be harmed.

Then came industrialisation. The slow, seasonal rhythm of land stewardship was replaced by the cold logic of profit. Soil was no longer seen as alive; it was seen as a machine, a resource to be exploited. The arrival of synthetic nitrogen and phosphorus fertilisers in the early 20th century promised to solve hunger and increase yields—and for a while, it did.

However, this abundance came with a price that we are only now beginning to fully understand. Synthetic fertilisers disrupt soil pH, break the delicate relationships between plant roots and mycorrhizal fungi, and reduce microbial diversity. The more they are used, the more they are needed, because soil organisms that once cycled nutrients naturally are killed or driven away. Healthy, living soil becomes dependent on its chemical fix, and without it, nothing grows.[32]

---

[32] SpringerLink, 'Fertiliser and Soil Degradation', (2022) <https://link.springer.com/article/10.1007/s11367-022-02078-1>.

And the damage does not stop in the fields. Excess nitrogen and phosphorus wash away into rivers, lakes, and seas, fuelling algal blooms that block sunlight and strip oxygen from the water as they decompose, a process called eutrophication. These oxygen-starved 'dead zones' now stretch across the world's waterways. Fish suffocate, aquatic plants rot, and entire food chains collapse. Lake Winnipeg, the Baltic Sea, and the Gulf of Mexico are all suffering due to agricultural runoff.[33]

Meanwhile, herbicides and pesticides coat our fields. They do not simply vanish after use. Many persist in soil, seeping into groundwater and killing not just the 'pests' they target but also earthworms, pollinators, and the soil bacteria that make nutrients available to plants. Over 95% of these chemicals drift outside their intended target zones, poisoning wildflowers, contaminating rivers, and further collapsing ecosystems.[34]

The monoculture trap

Planting only one crop in multiple fields strips the soil of specific nutrients, leaves it vulnerable to erosion, and invites pest and disease outbreaks on a catastrophic scale. To keep the system going, more fertiliser and more pesticides are used—a cycle that accelerates soil death.

And monocropping does not only harm the soil. It also kills diversity above ground. The long-running Park Grass experiment in the UK shows that fertilised, intensively managed grasslands have five times fewer wildflowers and 50% fewer pollinating insects than unfertilised ones. Bees are nine times more abundant on land that has never been chemically treated. Wildflowers are not just 'pretty'; they are food and medicine for pollinators, which are themselves keystone species in the web of life.[35]

Industrial farming has transformed once-living soil into a sterile, lifeless dust, reliant on chemicals for even the most basic fertility.

---

[33] US Geological Service, 'Nutrients and Eutrophication' (2021) <https://www.usgs.gov/mission-areas/water-resources/science/nutrients-and-eutrophication>.
[34] Beyond Pesticides, 'Biodiversity in Agriculture and Ecosystem Health' (2024) <https://beyondpesticides.org/dailynewsblog/2024/11/study-reinforces-importance-of-biodiversity-in-agriculture-and-ecosystem-health>.
[35] Beyond Pesticides, 'Monoculture in Crop Production and Pollinator Decline', (2019) <https://beyondpesticides.org/dailynewsblog/2019/07/monoculture-in-crop-production-contribute-to-biodiversity-loss-and-pollinator-decline>.

This is not farming in relationship with the land. This is mining, and the resource being mined is life itself.

It is now clear to science that soil is being degraded at an extremely high rate, and this has grave implications for the future. I agree with Rattan Lal, who rightly notes that:

> Soil degradation affects human nutrition and health through its adverse impacts on quantity and quality of food production. The decline in crop yields and agronomic production exacerbates food insecurity, which currently affects 854 million people globally. Low concentrations of protein and micronutrients (e.g., Zn, Fe, Se, B, I) aggravate malnutrition and hidden hunger, affecting 3.7 billion people, particularly children. Soil degradation reduces crop yields by increasing drought stress and susceptibility to elemental imbalance.[36]

The way back: Soil as sacred

Ecologically sound farming is not a romantic fantasy; it is the only viable path forward if we want to keep eating, breathing, and thriving on this Earth.

Natural fertilisers and compost: Returning animal manure, composted plant material, and even biochar to the soil builds fertility rather than depleting it.

Crop rotation and polyculture: Planting diverse species in sequence or in combination restores nutrients naturally and protects against pests without the use of chemicals.

Cover crops and green manures: Nitrogen-fixing plants, such as clover or vetch, can replenish soil nitrogen between harvests.

Agroforestry and hedgerows: Integrating trees, shrubs, and wild margins supports pollinators, prevents erosion, and creates wildlife corridors.

Low-till or no-till farming: Reducing soil disturbance protects fungal networks and increases soil carbon storage.

Local, small-scale systems: Transitioning away from large monocultures towards diverse, community-based agriculture enhances resilience and reconnects us to the land.

These practices are not 'alternative'. They are how we survive.

---

[36] R. Lal, 'Soil Degradation as a Reason for Inadequate Human Nutrition', *Food Security*, 1 (2009), 45–57.

The Pagan responsibility

In her book *Silent Spring*, published in 1962, Rachel Carson describes soil as the very origin of life, and the maintenance of its true nature is intimately related to the presence of living plants and animals.

In a restatement of her argument, published in 2015, she says that

> Soil is born of what we call the meeting of life and non-life, but is there really such a thing as non-life? How can there be, when life cannot spring from what is truly lifeless? We ourselves are shaped from the same elements as the soil and the rock—the iron in our blood is forged from ancient stars, the calcium in our bones from the Earth's crust—and yet we are called alive, while rock is dismissed as dead. How can that make sense? Perhaps it doesn't. Perhaps rock is simply life that moves at a different pace, a slower breath we cannot hear.
>
> Soil was forged aeons ago, when water and stone danced together under the Mother's gaze, long before we were anything more than a dream in her eye. Lichens were the first to stir the rock, eating it with secreted acids, breaking it grain by grain to draw out nutrients. Moss followed, making homes in tiny pockets of newborn soil. Slowly, life wove itself deeper into stone until the soil became what it is now: not just matter, but a living, breathing being.[37]

For us, as Pagans, witches, animists, for anyone who claims to honour the Earth, this should be more than an intellectual discussion. It should be at the heart of our practice. If we genuinely believe the land is sacred, if we call the soil 'our Mother', then every choice we make must reflect that reverence. What good is a beautifully crafted ritual if the food on your altar is grown in poisoned soil? What meaning does it have to call the land holy while supporting systems that are slowly killing it?

The soil is the first altar. The soil is the oldest cathedral. Every handful of living earth is a community, a mystery, and a prayer. If we wish to call ourselves Earth-honouring, we must step into right relationship with the soil—not just in our magic, but in the food we buy, the way we

---

[37] Rachel Carson, 'Silent Spring', in *Thinking About the Environment: Readings on Politics, Property and the Physical World*, ed. Matthew A. Cahn and Rory O'Brien (London: Routledge, 2015), pp. 150–55.

garden, the voices we raise against destructive farming, and the choices we make every single day.

Because if the soil dies, so does everything else. And if we save it, we save ourselves. 'Dust to dust' is no soft comfort for funerals; it is law. Nothing ends; everything folds back into itself. The black skin of the Earth teems with quiet labourers, worms and insects shredding the fallen, bacteria whispering their endless alchemy, fungi threading their bodies together. A single teaspoon of living soil holds more life than most forests. The rock we call lifeless feeds it; the so-called dead sustains the living.

These are the Black Paths, the memory of everything that has ever lived and died, twisting and knotting beneath the damp moss and leaf litter.

The Silver Highways

It is a lie to think of trees as lonely sentinels, standing apart, raising their arms to the sky in isolation. The truth hums, no, thrums, beneath our feet. The Silver Highways run deep in the black earth, weaving root to root, heart to heart. They do not whisper in slow, ancient syllables as we once believed; they flash like thought, like a pulse of living electricity.

A wounded tree can cry out along these fungal roads, and far away, a sapling stiffens, leaves already flooding with bitter poisons before the predator even arrives. The latest research indicates that these networks transmit warning signals in a manner that closely mimics the electrical pulses of an animal's nervous system. The forest feels. The forest reacts. The Silver Highways are not just fungal threads—they are nerves of the Earth, alive with information.[38]

And they are not only messengers; they are lifelines. Healthy forests feed each other through these silver threads, sending water, carbon, and sugars to the sick, even nursing fallen trunks, keeping them alive for decades so they may send up shoots again. This is not competition—it is community, a living pact. Mother trees, ancient and vast, push sustenance into the network for shaded saplings that would otherwise wither. The forest shares its memory, its strength, its will to survive.

---

[38] Matteo Buffi and others, 'Electrical Signaling in Fungi: Past and Present Challenges', *FEMS Microbiology Reviews*, 49 (2025), fuaf009.

In healthy soil, this conversation sings. The black paths smell of rot turned to promise, rich with the alchemy of decay and renewal. The Silver Highways lace them with glomalin, that miraculous fungal glue binding chaos into order, holding water as lungs hold breath. Microbes bloom, roots reach like dancers for their partners, and the electric pulse of tree-thought moves through the ground in quicksilver flashes. Even seeds expect this partnership; they are born coded to reach for the fungal weave as soon as they crack their shells.[39]

But when we break the black paths, silence falls. Starved soil is a body without veins; the nerves go dead. Roots claw at dust and find almost nothing. Microbes starve. The silver threads shrivel. Above, the green cathedrals weaken, their vaulting spires thinning to brittle grey until they stand like bones, hollowed and mute.

The forest is not a collection of individuals; it is a single breathing being. The Black Paths are its lifeblood and memory; the Silver Highways are its nerves, carrying thought and feeling through a body vast beyond our imagining. The swaying canopies, those vaulted green cathedrals, are its song, an old hymn of life sung for thousands of years.

Walking the Black Paths

The Black Paths do not open easily. They are not the soft whisper of leaves or the easy greenness of trees; they are older, slower, and stranger than anything we are used to. If you would walk them, you must ask properly and move with respect, for this is the hidden machinery of life and death, and it has no time for careless footsteps.

Preparation

The night before, keep yourself light. Eat plainly—roots, grains, or nothing at all if you can bear it. Speak little. Let your mind quieten. Clean your space, sweeping as though each stroke of the broom clears not just dust, but the restless thoughts you carry.

Tie a black cord with three knots. As you bind each one, whisper:

> *When I wander far, this will bring me home.*

---

[39] M.J. Richardson, 'Seed Mycology', *Mycological Research*, 100 (1996), 385–92.

Keep it close to you; it is your lifeline to the waking world.

Set your working space simply. This is not a place of grandeur, but of truth.

This meditation should be done outside, on the land you are making communion with.

A living token, leaf, feather, or bone, an echo of the world above.

One candle, small, steady, beeswax if you have it.

A vessel of water, for the return.

A small seed, piece of bread, or a pinch of herbs, as an offering to the unseen world.

Sit on the soil and place your hands onto it. That is where you will enter.

Opening — Calling the path

Feeling the ground beneath you.

Speak softly, letting the words fall into the earth more than into the air:

> *I sit at the crossroads of the World.*
> *Out of time and out of place*
> *Beneath me run the Silver threads,*
> *Tapping rhythm into space*
> *Above me shine Cathedrals Green*
> *Fed by the Sun from whom light streams*
> *Between them, I walk the Paths of deepest Black.*
> *Where Life and Death entwine and speak.*

Breathe until you feel yourself settle, as if roots from your own body are reaching down into the dark earth.

The ask — Whispering into the dark

The Black Paths do not answer idle curiosity. Be clear and honest in what you seek. Whisper your question or purpose into the soil:

> *Teach me*
> *Show me*

Do not ask for power or gifts; ask for truth. Breathe into the earth below you, sending your intention into the hidden threads.

The journey—stepping onto the Black Paths

Close your eyes. Begin whatever carries you inward—breathwork, drumming, silence, or the path you have chosen.

Focus on the feel of the earth under your hands. Feel yourself slipping through it, following the threads beneath. The world above will thin; sounds will change. There may be a clicking, a distant hum, or the sense of a slow, grinding movement; this is the deep speech of the Black Paths.

Do not try to shape what you see or feel. The Black Paths will show you what they choose. Take it as it comes. You are a guest here.

The return—calling yourself back

When the pull loosens or you feel it is time, pick your hands up from the surface of the Earth and say

> *I am released from the Paths*
> *The Threads will let me go*
> *I stand again at the Crossroads of the World.*

Drink from the vessel of water. Feel its weight pulling you fully back into your body. Untie or hold your knotted cord and speak:

> *I return. I am whole.*

Closing and integration

Leave your offering on the earth as thanks.

Write down what you remember immediately, shapes, sounds, feelings, even if they make no sense now. Eat something grounding, such as bread, roots, or seeds.

And above all, thank the soil. Always thank the soil.

As with all things, the soil you work with has a *personhood*. I have worked with the soil in my area and developed a good understanding of it through my years of practice.

My magical practice centres on the land where I live. I am fortunate to live in a rural area of Wales, 15 minutes away from the coast. I have a relationship with two areas of land around my house, and I am in the process of forming a relationship with the land in a place called Southerdown on the coast of South Wales.

Meditations on the soil in my garden and the wood behind my house

I walk this land as a traveller, not an owner. Every step is a negotiation, every breath an exchange. The spirits here know me; some tolerate me, some welcome me, and a few still test me. This is their place as much as it is mine, and I tread with care.

The soil is clay, sticky, red, stubborn. It clings to me, pulls at my boots, makes me work for every seed I plant. I do not fight it; I feed it. Chicken manure, compost, time. It rewards my patience with vegetables, ferns, and flowers, but only when I show respect. This soil is old, proud, and slow to give its trust.

The honey fungus lives here, weaving its silver threads through the Black Paths beneath me. It is no easy ally; it is a teacher who does not soften its lessons. When I sit with it, in meditation or simply in silence, its rhythm clicks against the edges of my mind, patient and almost alien. It does not care for me, but it lets me listen. In its endless work of decay and rebirth, I learn something of the world's true pace, of how life feeds life in the dark places.

The coal seams beneath whisper of old scars, of the time when humans clawed too deeply. The stream that runs rusty at the back of the garden hums with that memory, but still, the trees grow thick along its edge. Hazel laughs, quick and darting. Holly watches me with her white-boned pride. The oak stands slow and certain, letting me rest my back against him. My baby yews, planted as honour to the God Tree, reach down cautious fingers into the earth, beginning their long conversations with the land.

The garden is never silent. The beetles run their hidden markets, the spiders weave tight little stories between branches, and the worms carve their endless roads through the Black Paths. When I lie down in the soil, I feel them moving all around me, unbothered by my weight, busy with their own work. This is not my kingdom; it is their world, and I am only a guest.

But I am a welcome guest now. The land knows me. It has held me as I cried, as I birthed my daughter, as I buried my dogs in its arms. It has watched me scatter seeds and dig graves, chase cats and chase dreams.

Sometimes, when the rain has soaked it to softness, I sink ankle-deep, boots pulled at with that wonderful, greedy squelch, and I laugh like a child. Other times I lie still, pressing my cheek to the soil, and feel it rising around me, holding me close. The spirits press in then, not with words, but with presence, the soft murmur of the green people, the sighing hush of the winds, the quiet hum of roots and threads.

This is my sanctuary as much as any circle I cast. Here, in this patch of clay and iron and leaf litter, I walk between the spirits. I trade breath for breath, touch for touch. I am part of the web, not above it.

This is my sacred ground. My crossroads. My teacher. My home.

Meditation on the soil in the field

I know a field not far from my house. I can walk there easily, and I often do. Its soil is strange, layered over the bones of the old mining cottages that once stood here. Now they are nothing but memory—stone ground back into earth, wood rotted to soil. This field reminds me that all our marks, all our clever human work, vanish in the end. Time eats everything. Nature always wins, and perhaps that is right. Perhaps we do not deserve permanence if we cannot learn to live in harmony with her.

The soil is wet and boggy, sometimes full of cows. Because of this richness, in late spring, pale pink orchids rise here, their delicate dresses trembling against the wind, defiant and unafraid. In the centre of the field stands an old thorn tree, gnarled and steady, her roots holding firm in the soft ground. Beside her lies the body of an oak struck down by lightning long ago. I often wonder if she misses him.

I walk here in autumn when the light is low and the ground smells of water and leaf-mulch. I look for allies, for the teacher-beings that open the mind to the divine, those plants that show us how to live better on this green earth. But this is not their place. Here, the soil grows grasses and reeds more than it grows teachers. Yet it remains sacred all the same.

For my dog, Douglas, this field is his wonderland. He is old now, no longer crashing through puddles, no longer chasing the wind. He walks slowly, nose to the ground, drinking in every smell, taking his time. He still eats cow pats when I am not quick enough—*"Douglas, put that down."* I laugh, even as I scold him.

He will not be with me forever. I know it, and I think he knows it, too. Lately, he walks differently, pausing longer, sniffing the reeds and grasses as if choosing something. I wonder if he is searching for the place he will lie when he is done running, the spot where he will finally rest and give himself back to the soil. If so, I trust his choice. This is good earth to return to.

This field may not teach me in the way I ask, but it teaches me all the same. It teaches impermanence, patience, and acceptance. It whispers that life and death are just two steps on the same path, and even our love, when it fades from breath, will remain in the soil, in the grasses, in the roots of the thorn tree.

This field, too, is home.

Meditations on a newly forming relationship with the coast

This soil is nothing like home. It is dry, sandy, loose beneath my fingers, and the trees here all lean into the endless wind. The land itself shouts—its voice full of water crashing and stone breaking. Beneath the sand it hides fossils, small curled memories of lives far older than any of ours. The cliffs rise sheer from the sea, the bones of the Earth laid bare, layer on layer, age upon age. Humans are nothing here. We are new growth, soft and temporary, clinging to the edges of something ancient.

The trees twist themselves into strange serpentine shapes, reaching for the light as if they have been fighting it for centuries. We sit among them, and before we begin, we ask permission to work. The spirit answers. There is a feeling of relief, almost joy, as if the land has been waiting for someone to notice, someone to care. It does not ask for much, only presence, an offering of being seen, of being heard.

We walk the edges of the land, and it shows us its wounds. Rubbish left in heaps, the burnt scars of old fires, sanitary towels discarded and forgotten. The plants fight to live here—blackcurrants taste of salt instead of sweetness; grapevines shrivel but cling stubbornly to their walls; roses sprawl wild and defiant, throwing thorns into our fingers when we try to raise them for a sniff. *You may not take us,* they seem to say. *We are here for the wind, not for you.*

There is laughter and shouting from a nearby garden. People playing football, unaware or unconcerned that they are being watched. Not by us, but by the green people, by the ones who live deep in the soil and the rock, the ones older than the sea cliffs themselves.

We work quietly, offering what service we can to the land, to the sand, to the salt-worn plants. We give it presence. We breathe with it, and it breathes with us. It will take time, but I know one day it will speak, tell us its stories of water and stone, of wreckers and storms and the long memory of fossils.

For now, we wait. Breath held. Listening.

Magical work with the soil

Magical working on a local nature reserve to prevent building on the ground

The local council made plans to build residences in an area near us in Wales, which is a nature reserve. This place is fantastic as it provides for birds, including herons, swans, swifts and martins. It features newts, snakes, mushrooms, and a variety of trees. It truly is an oasis of nature. To protect the land from development, we created a *'golem'* from the sand and mud of the land. We invoked energy into the being and then set it to guard over the land.

One of my coveners made a land pact in this place to service and feed the *Guardian Spirit*. Since we performed the magical work, the plans have now been shelved, and the site was designated as one of *'Scientific Interest'*. Here, the personality of the land is one of great power; the guardian has been accepted, and this is our go-to place for mushrooms and plants that live by the body of fresh water. The land undergoes significant changes with the passage of the seasons. It is, therefore, an ideal place to understand time, the seasons, and our work with the ecosystem and the energies of our green friends.

Soil magic

In many cultures, this work has long been referred to as dirt magic, the art of burying charms, binding spells into the ground, or letting the earth hold what we no longer wish to carry. Across time and tradition, soil has been a place of release, transformation, and banishment. We give our unwanted things to the dirt because we know it will take them, break them apart, and return them to the great turning wheel.

But I cannot call it dirt. Dirt is what we wipe from our boots; dirt is a dead thing. The truth is that soil is alive, and to work with it effectively, we must see it as a living entity, not just a tool. Soil is an ally, a partner in the work, and when we are in relationship with it, it aids us as surely as any plant spirit or stone. This is not dirt magic; it is soil magic.

Growing spells

Using a growing spell means you plant your spell within a plant seed. If you want a quick result, you can use a fast grower like cress. If you want to employ esoteric *sympathies*, then you should also use astrological correspondences in your work. I tend to follow Culpeper for my astrological correspondences, but you may want to use other sources. I also add my own personal magical take on them because of my close work with the plants. For this reason, I tend to narrow it down to a zodiac sign and a planet, rather than just the planet, such as Culpeper recommends. For example, I work with mugwort as Venus in Cancer. She possesses both aspects of Venus, as reflected in her work with the female reproductive system and her role as an Opener of the Way in the psychic realm. She also displays characteristic traits of the Moon.

Thus, as a quick rule of thumb, use *sympathies* and begin by planting a seed with a spell in the soil, and as the seed grows, your spell is released and will increase. However, if your plant does not grow well, this is a good indicator that your magic is not viable.

Banishing spells

Burying things in Earth is also a good way of moving them on; for millennia, we have buried our dead to allow their spirits to move onward. In the magical tradition, burying items in soil can also be an excellent way to banish them.

Method: write a word or draw a sigil representing what you must banish on a rock, paper, or bay leaf. You could banish an emotion you no longer want to feel, a person in your life who is bothering you, a habit you want to rid yourself off, or a fear that is holding you back. So, on a waning moon, take the object into your hands and visualise the energy of whatever you're trying to banish moving from your body, out of your hands and directly into the object. Bury the thing in the soil and leave it there.

Soil—magically used as a binding object

Soil from your own property can be used as an ingredient when you want to protect your home and the people in it.

Soil from the property of an enemy can be used to bind or put a *hex* on them.

Soil from a bank can be used for prosperity work.

Soil from a church can be used in spiritual matters, such as prayers and petitions.

Soil from an ancestral home may be used to help connect with the spirits of your own ancestors or strengthen your connection to the past.

Soil from a place you want to live can be used to pull that place towards you.

Earth-based divination systems

The origins of geomancy

The earliest mention of the word (*geōmanteía*) in Greek is associated with Archimedes (287–212 BC), who reportedly drew *geomantic figures* in the sand during the siege of Syracuse to determine the outcome of the situation. However, it remains unclear when exactly geomantic figures first appeared, even though the drawing of magical formulae in the sand is probably very ancient. It seems likely that the *praxis* probably arose in North Africa, although the late-Greek term *geōmanteía* translates rather literally as 'earth divination'.

It is perhaps better to examine the later spread of geomancy that resulted from the expansion of Islamic thought throughout the North and the West. For example, in West Africa, the praxis is known more commonly as *ifa*, and in nearby Dahomey as *fa*. In addition, the art of geomancy also travelled down the Red Sea to Madagascar, where it became known as *sikidy*. It eventually travelled to northern Spain, where it became more traditionally known in Europe simply as *geomancy*. When America was finally reached by Europeans and African slaves from the African continent, various forms of the *arte* soon began to flourish.

Owing to its widespread origins, the methods of producing figures and their meanings differed from culture to culture. However, the original Arabic custom was to make marks in the sand, and the divination was simply referred to as *raml*, meaning 'sand'. This was most likely based on the late-Byzantine Greek term *rhamplion* or *rabolion*. It was this method that resulted in the practice eventually being called 'geomancy'. In Europe, even though pen and paper replaced the sand tray and stick/wand, the term itself would subsequently remain.

In comparison, the more traditional *fa and ifa* approach involves using 16 'palm nuts' picked up quickly with the right hand. As the nuts are large, some inevitably drop, and the figure is produced by counting the remaining nuts that are left in the hand. Another alternative uses a 'divining chain', which is made up of a length of chain with *cowrie shells*, and these are used to generate the magical divining figures.[40]

Thus, although, as we have seen, the earliest mention of *geōmanteía* is found in relation to Archimedes, the nature of the process he used cannot satisfactorily be established other than with the meaning *'divination by earth'*. As a divinatory system, the term implies one of Greek or perhaps even Indian tradition, especially one where a numerological system is based on odds and evens. Yet, this process can also be found in the Hebraic use of the *urim* and *thummim* (the yes-and-no stones/bones), which first appear in Exodus 28:30, where they are named for inclusion on the breastplate (the *ephod*) worn by Aaron in the holy place. It is perhaps no surprise that this type of divinatory process would eventually make its way northwards and to Spain, where it became more traditionally known in Europe as 'geomancy'.

Europeans and African slaves reaching North America brought their own forms of geomancy. However, the methods of producing the figures and assigning meanings differed for each culture. For example, the Arabic custom was to make marks in the sand, and the name assigned was thus, *raml*, sand, hence, geomancy (lit. 'reading earth'). In Europe, a pen and paper soon replaced the sand tray, stick, or wand.

The workings of geomancy

A geomantic figure is made up of four lines of either one or two marks. In the traditional way of working from *raml*, the diviner, while concentrating on the question at hand, produces a line of marks or dots in the sand with a stick. The number of marks in the line is then counted. An odd number produces one mark for the line, while an even number produces two marks. This operation is repeated four times to create a single figure consisting of four lines. In the system of *fa and ifa* the diviner counts the number of remaining palm nuts in the left hand: an odd number yields two marks and an even number yields one.

In *raml* and *sikidy*, there were further applications for a more involved Medical Geomantic analysis, and this praxis can also be seen in the European model. The geomantic figures were associated as follows: the head *Laetitia*; the throat *Rubeus*; the right shoulder *Puella*; the

---

[40] Cf. Stephen Skinner's *Terrestrial Astrology: Divination by Geomancy* (London: Routledge, 1986).

left shoulder *Puer*; the chest (heart?) *Carcer*; the right side of the ribcage *Coniunctio*; the left side of the ribcage *Populus*; the solar plexus *Albus*; the stomach *Via*; the right hand *Amissio*; the left hand *Acquisitio*; the right thigh *Fortuna Major*; the left thigh *Fortuna Minor*; the genitals *Tristitia* (*Laetitia*?); the right foot *Cauda Draconis*; and the left foot *Caput Draconis*.

As we can see from the list, the terms and sympathies can be recognised from their Latin terms and in this way can also be linked through *astrosophy* and healing. The geomantic figures can then be used in a more magic way, and as such by creating talismans. The talismans can then be charged and placed on the specific area of the body that is affected by disease.

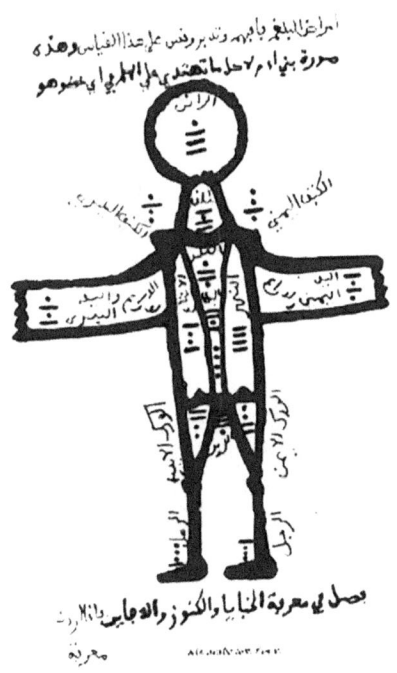

Western geomancy

In European geomancy, four figures called the *Mothers* (*matres*) are produced using the random marks method, and from these, further figures are calculated. The divination can be made directly from the figures themselves or by using a system of correspondences translated onto an astrological chart and then interpreted according to astrological rules.

The 16 geomantic figures

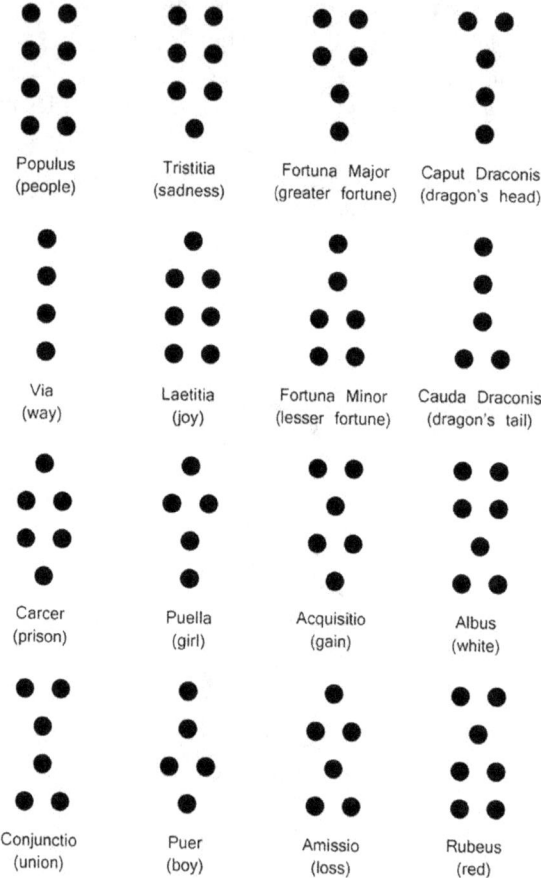

So, how do I use it?

In its simplest form, geomancy can be described as earth divination. It is based on gaining insight into present and future events, obtained through subconscious activation and conscious action, by observing certain combinations of patterns made in the Earth (or on paper) by the diviner. It is therefore a praxis that allows intuition to act with 'the spirits of the earth', enabling control over the movement of the wand or pencil.[41]

---

[41] Ibid.

Casting the points

The first step is usually called 'casting' or 'sowing' the points. In this step, the geomancer, while focusing on a specific question (to be answered), draws 16 lines of points, from right to left. Some treatises advise that the geomancer must cleanse and then pray before casting the points. Personally, I would recommend an invocation to the spirits of the geomantic figures in general, as well as a prayer to the spirits of the Earth who control the entire system and will aid in the operation. No effort should be made by the practitioner as they mark the points in the paper or sand. However, the geomancer is usually advised to make at least 12. Casting the points is a critical process in geomancy; if the geomancer miscasts the points, the tableau will be invalid. The example shows the first four lines (right to left, created by the geomancer). When each line is added, it either creates an even or odd number. The one below creates the figure known as *acquisitio*, which is created top down (2, 1, 2, 1).

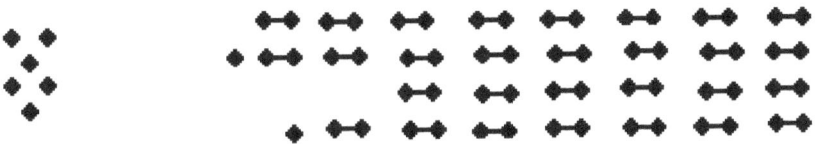

In the traditional way of working from *raml*, the diviner, while concentrating on the 'question in hand', will produce with an almost 'unconscious' series of marks a series of 'dots' in each of the four lines—these marks are pocked in the sand. The number of marks in each of the four lines are then counted. For example, an odd number (say 3 dots) produces 1 mark for the line, and an even number (say 6) produces 2 marks. The operation is repeated four times to create one figure with four lines (say for example 7, 4, 3, 2, which produces a geomantic figure consisting of odd (7), even (4) odd (3), even (2)— the new figure ends up with 1, 2, 1, 2 dots. In the *fa and ifa praxis*, the diviner counts the remaining palm nuts in the left hand. An odd number will yield two marks and even number one, so you create the following.

Automatic: the four originally poked lines in the sand, or those made by dotting on paper—for example, in this case, it will return the geomantic figure known as *amissio*. You must focus your intention, but remember not to count the dots as you progress.

> The 7 (odd) produce   .
> The 4 (even) produce   ..
> The 3 (odd) produce   .
> The 2 (even) produce   ..

The geomantic figure of *amissio* is therefore produced:

Working with geomantic figures in astrological magic

I started working with geomancy several years ago, initially using it as a magical tool rather than a divination tool. However, once you begin to accept that geomantic figures can be used as *doorways and access points* as astrological *planetary energies,* you then have an unlimited potential for an *Earth-based magical system.* So, if you want to pursue this idea further, I advise you to seek out Dr Al Cummins who has some excellent courses on the subject.[42]

Developing a more practical use for the geomantic figures

In one of our magical workings, we utilised the geomantic figure Carcer to prevent ash dieback disease from killing the trees in our area.

The ritual

The ritual is a *binding ritual,* which is best done during a waning moon. We used bindweed, a Saturnian plant, and the geomantic figure *Carcer,* which, as the Latin term implies, means 'binding' or, more literally, 'prison'.

---

[42] http://www.alexandercummins.com/#classbundles-section

Carcer has *Active fire*, which will kill the fungus and sterilise the tree. *Passive water and air*—this fungus is transmitted through water and air, and we wanted to prevent that, so passivity is a good sign to utilise. It also features *Active earth*, so we are enclosing it in the earth, hence the use of a salt ring.

Let us offer an outline of the ritual for the protection of the ash trees.

An invocation to Ash

Connecting tree
Broad-branched Nuin
I call your name and summon you forth
Come green clothed healer man
Stang of the world
Accept my aid, and aid me in this working.

Finding of the fungus

[as you are saying this bind the dead ash wood with bindweed]:
Hymenoscyphus fraxineus[43]

Tree killer
I bind you from harm
I bind you from harm
I bind you from harm
Stop your invasion
Stop your invasion
Stop your invasion
By Bindweed, child of Saturn, I bind you
By Blackthorn, child of Mars, I hold you
By Carcer sign of Saturn, I imprison you

Plant aiding plant
My spirit the binder your spirit the bound
I bind you from the harm you do
Without malice with love, I bind you

For the good of the Great Ash
In healing the healer, I heal the world.

---

[43] The Latin name of the ash dieback fungus.

As you recite the above (or perhaps something of your own design), actively:

*Bind the bindweed around tightly
And then move the blackthorn, then walk widdershins around your tree/*

Take a piece of the ash with the fungus on it (a piece without leaves from the top of the plant) and place it within the figure of Carcer:

Surround ash and fungus with candles to make the figure of Carcer, then surround the whole with a circle of salt, as seen on the accompanying image.

DragonOak has also used geomancy in a novel way by combining it with *Morris Dancing*. This ritual was used to raise the energies and inspire us to move forward with our Green agenda, replacing the more destructive technologies currently in use. I now invite the reader to follow *An Ecological Magical Performance* produced by our good friend and practitioner Lucy Greenwood.

Ritual incorporating a traditional Border Morris dance [by Lucy Greenwood]

I would like to thank Sian for including the following ritual in the book. For many years, I had danced in a Border Morris side. I felt that when I danced with others, the energy created could be used for a more magical purpose, especially when all those involved produced a powerful and collective intent. Sian has been my mentor in the craft for many years, and I received my training with DragonOak, where I have remained within the coven for the better part of the last decade.

The following ritual primarily exists due to DragonOak's attitude towards magic. We all research and read about the subject of the craft, and we are also encouraged to try things out on a more practical level; sometimes things work, and sometimes they don't. Fellow coveners are encouraged to share their thoughts and opinions and from these discussions, we develop and employ rituals together. I therefore thank them for their contributions to this ritual and for the time and energy they invested in performing it.

Equipment: Hazel stick (one metre in length for each person participating in the dance);

Strong voice: With true and focused intent;

Time and place: Must be chosen in advance, depending on ritual focus.

Purification and robes: To be used in accordance.

Offerings: I suggest scattering seeds for the wildlife and perhaps making incense from the local area's bark and herbs. Only make as much as you need. (NB: Do not use yew or elder.) We also offer our service to the land here regularly, by gardening, picking litter, and weeding.

Intention: To facilitate a more positive way of moving forward as a civilisation with the well-being of all people both human and non-human in mind.

Purpose: We understand that economic growth will continue, but that that growth *must be Green*. We must find the value in renewable resources. We must clean up. This is the end of the 'industrial' and the beginning of the *Ecological Revolution*.

Please note that we do not 'cast a circle' here as we want the guardians, spirits of place, and winds to aid in our *praxis*.

Call guardians: these can be called from your own *praxis*.

Our own DragonOak variation:

> We call upon Euros, the eastern wind, to spread our cause across the land.
>
> We call upon Notos, the southern wind, to spread our cause across the land.
>
> We call upon Zephyrus, the western wind, to spread our cause across the land.
>
> We call upon Boreas, the northern wind, to spread our cause throughout the land.
>
> We call on the spirit of the land to aid us in our work, to accept us as supplicants and workers of magic.
>
> I call upon the earth that I stand, ancient earth, earth older that no person can speak of your true form and spirit: Ged
>
> Terra
>
> Erda
>
> Gaia
>
> The foundation and vessel for all life, we honour you and are thankful to you for our lives. We recognise the damage done by our hands and wish to implement change so that we call all to see that we are profoundly united with all your living bodies. Animals, plants, soil, rocks, rivers, seas and air.
>
> Bring clarity to the decision makers, open their eyes and hearts to the dangers of unsustainable growth, not only to the human race but all of Earth and those who dwell upon her.
>
> We do not call for stasis; we call for new technology that does not pollute but instead cleans up pollution already caused. We call for solutions that do not just benefit developed countries at the expense of others, instead developing all countries with Green and new technologies.
>
> We call for a growth that is Green, for the decision makers to find economic value in the ecological wellbeing of all.

The dance

The dance allows the ritual participants to join together to form specific geomantic figures. The 'figures' must represent the specific intention of the magical working. If you want to work a solitary praxis, I suggest that you put the figures on the ground using Markers before you start, to make it easier for yourself.

For the group work, a traditional hazel stick is held in each of the practitioner's right hands; sticks are to be held out towards the middle of the 'first position', and practitioners start to dance in a circle—*deosil*.

The first four dancers get into position. There will be a 'chorus' between making each of the geomantic figures and where the repetition and energy-building aspect of the magical work creates the dynamic.

The circle dancing will be referred to as 'dancing the round'. When we move in the dance, we refer to this as 'stepping'. Stepping is therefore a little like skipping, but instead of kicking your leg behind when you skip, you kick it out in front of you (you can find several online tutorials of the *praxis*).

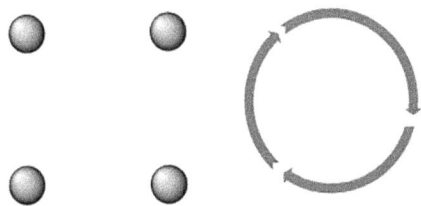

The dance is done to a simple 4/4 beat. It is useful to have someone drumming, or alternatively, you can just keep time in your head. Each 'figure' takes eight beats to complete, therefore if the dance calls for something to be done three times, this implies that you should have three lots of eight.

Now step around 'in a circle' three times while chanting: "*Growth in the Green*"

Dancers will now form the first 'geomantic figure'.

Now—Step forward for four counts and backwards for four counts in a line.

The first geomantic figure to be formed is that of *Via*:

○

○

○

○

The geomantic figure *Via*, or 'way', has all elements active, which reveals its dynamic nature; its linear form resembles a road or way, as the movement that is undertaken will project time and circumstance. This will transform every other figure and hence forces them into a series of opposites. *Via* therefore holds the possibility for complete change, upheaval, and/or reversal.

We chose this particular figure as the 'change' we needed to manifest a complete reversal in the current status quo. We must then magically manipulate the world's ecology, and this can only be achieved by putting it before our own profit and indeed lifestyle.

Now with gusto chant the following twice: Via (ho) Via (hey) Via (ho) Via (way).

On the third chant line up to form *Via* on last *ho* sticks on the ground, and on the last (way) sticks up in the air.

Dancers then return to the round and chant the chorus and x2 more dancers join.

Chant: "Growth in the Green"

The second figure formed is *Coniunctio*:

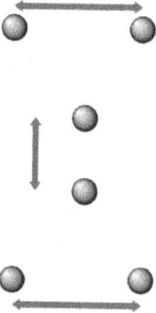

WORKING WITH THE LAND SPIRITS   105

The Latin term *coniunctio* means 'conjunction/joining', but can also be translated as 'assembly' or more simply 'meeting'. If you draw (arrowed) lines on the figure you will see that the even points literally 'join hands' and form a bridge. We therefore chose this particular *geomantic figure* as it suggests the 'coming together' of all concerned and by way of re-establishing a moral balance between governments and businesses, and in such a way can demonstrate how the planet is affected by ecological disaster. Our magical purpose is therefore presented through a coming together in a truly ritualised form and energy, and this is presented through *coniunctio*—otherwise, all will end in ecological disaster.

This time, dancers will face each other as shown in the diagram, and they will 'stick' with each other. Each dancer will have an opposite dancer facing each other: you step forward, and swap places and then swap back.

Chant Coniunctio (Hey—sticks in the air) *Coniunctio* (Ho—sticks on the ground) x3. The *heys* and *hos* are said on every fourth beat.

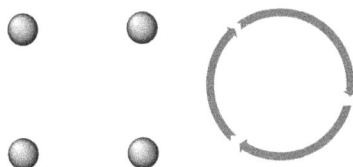

Dance the round with the same number of dancers circling *deosil*.

Chant: "Growth in the Green"
The third figure is *Acquisitio*:

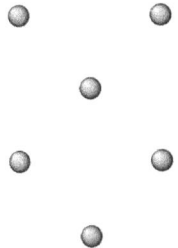

This geomantic figure is known as *acquisitio*, otherwise *'gain'*. The elemental structure is with air and earth active. The overall figure mimics that of a bag or bowl held upright with objects falling into it. The figure also demonstrates the combination of intellect and communication with material reality and where manifestation can bring about the attainment of a goal, money, or objects. In this respect, *acquisitio* corresponds to Jupiter, and where we want 'all' to benefit through the control of a much greener industry—we therefore magically activate and induce the energy of becoming a better and cleaner form through *acquisitio* in general.

Step forward and back three times chanting: "*Acquisitio Hey Acquisitio Ho*" x3—on each *hey* and *ho* sticks to the air and sticks to the ground respectively. Dance the round with x1 more dancer joining.

Chant: "*Growth in the Green*"

Figure four is *Albus*:

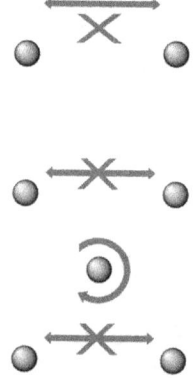

Albus is Latin for 'white'. The figure resembles an upright glass or goblet. Astrologically, *albus* is associated with Gemini and Mercury; even though its inner element is water, its outer element is air. It represents peace, wisdom, and purity. It benefits beginnings and profit, or any situation requiring careful and deliberate planning.

We chose *albus* for its associations with communication and as a guide towards better planning processes by businesses and governments in order to ensure that they consider the environment when undertaking planning the way forward.

Now, split into x2 groups, (4) square, (3) triangle to make the 4th position *Albus*. Each pair will cross paths and clash sticks in the centre. Represented by the X in the diagram. The solo dancer will spin in the spot. Chant: *Albus* (Hey) *Albus* (ho) x3 with the sticks to the ground and to the air. Everyone now dances in the round:

Chant: "*Growth in the Green*" but faster and faster with sticks raised in the middle until it feels right to release the power and drop the sticks to the ground.

NB Because the sticks are literally thrown to the ground at random—their new form can be 'read' as another separate geomantic figure and which can provide an overall divination—this can then be read in order to conclude and to see if the main ritual that was performed will have a positive outcome.

Conclusion

For the work to be successful, it requires much energy and movement, so be sure that everybody in your group can perform the dance in an active fashion. However, those who are unable to participate more fully in the dance aspect of the work can still nonetheless contribute significantly by keeping time—especially with a drum or other instrument—and by contributing to the chants, and even by calling out the figures and giving instructions on when to move into position as at first this can be somewhat confusing. Practising the movements and dance helps greatly before actually setting the ritual and initial intention before doing the work properly.

If you are not working in a group, you could mark out these figures on the ground with pebbles in soil or sand and dance and chant around them, building up the energy and sending it out all the same. This experimental work welcomes modifications, variations and other magical additions, depending on your needs.

In the work, we used four geomantic figures that will not only bring about change to the current mindset of prioritising profit above all else. The rite communicates 'change', the 'bringing of ideas together', whereby we project the group's collective mind and energies towards acquiring more *green technology*, which will impact the planet's present state.

We are not seeking wholesale change or a Luddite reversal here; we are accounting for the profit-driven motives of most governments and businesses.

Magic works best when it flows with the energy of the situation. We will not change the mindset of business and governments from their capitalistic ideals overnight. For this reason, we agree that it will take time and hard work to move the mountain, but we must move it; otherwise, we won't survive.

Results from this work—Note for 2nd edition, Sian Sibley (July 2025)

In the two years since we began this work, we have seen a marked increase in the amount of Green industry, specifically in the area around the location where we performed this ritual:

- Three new wind farms have been established.
- There is a new Green park where people can hire bikes and arrange car shares and busses.
- A new hydrogen manufacturing plant is being founded.
- Two new solar parks are being built.

Overall, I would say that the ritual was an outstanding success.

Eco magic to protect a woodland area

There are moments when a place cries out for help.

A woodland near me had begun to wear its grief openly—litter scattered like wounds, undergrowth trampled and torn, the quiet places broken by fires and voices. Some had taken shelter beneath the trees, their need written in ash and foil and broken glass, and though compassion ached in me for them, the land's spirit was dimming. I could feel it—that slow retreat of presence, the soft recoil of something once bright and breathing. Words had failed, reason too; the polite signs asking for care hung like forgotten prayers. The forest needed protection, not permission.

So I made a guardian. Not of clay, as in the old stories, but of porcelain—a small doll I had kept for many years. Dolls are strange creatures; they hold a tension between innocence and haunting, between child and ancestor. They are vessels waiting to be filled. Upon her pale body I inscribed sigils, each drawn from a sentence of intent: *you shall not enter, you are unwelcome, you will feel uneasy, protect the land.* These were not curses but boundaries—lines redrawn at the request of the Earth herself, a reassertion of sacred ground.

I passed the doll through smoke, breathed life into her hollow chest, and placed her in the hands of another—the woman for whom I had crafted her. For the bond between guardian and keeper must be direct, a thread tied heart to soil. She walked the woodland paths with the little effigy cradled against her, whispering her name, her purpose, her promise. At the heart of the forest she found an old oak, its arms wide as wings, and there she laid the doll to rest. Words were spoken, quiet but sure—a charge for guardianship, a vow that this spirit would stand watch with the trees, deterring harm and holding the boundaries of peace.

Now the magic breathes. The porcelain guardian lies hidden in leaf and shadow, unseen yet awake, her awareness interwoven with the forest's dreaming. The air feels different now—still, listening, cautious. We do not yet know what will come of it, what ripples will move through soil and season, but something has shifted. The work is alive.

And so we wait, with bated breath, to see how the woodland will answer—how she will breathe again beneath the quiet vigilance of her small, enchanted sentinel.

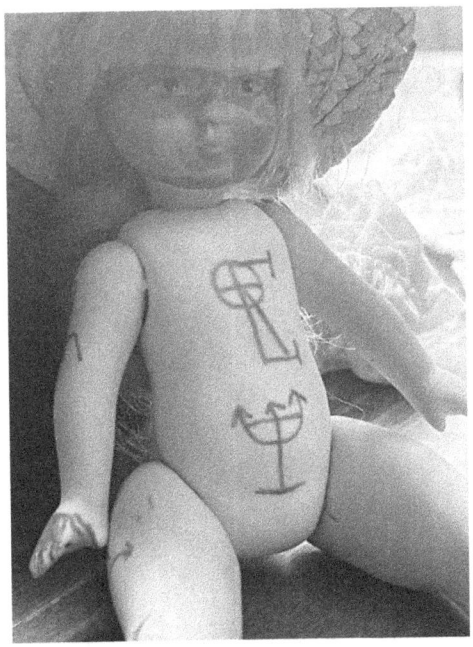

PART THREE

GREEN CATHEDRALS: THE MAGICAL TREES

CHAPTER FIVE

Working with Charubel's plant sigils

Working magically with trees takes a great deal of personal effort. It is long-term work which, with any luck, will outlive me. The trees I work with are at the back of my house, so they constitute my *genius loci* and are people who greatly help me. This book, in part, discusses my experiences and magic with these magnificent beings. I will include some herbal information, lore, and magical information with each plate that follows. I am hopeful that when others see what I have been doing, they too will start to work effectively with these amazing green people.

I follow in the grand tradition of Charubel, a Welsh shaman whose work I share in the following pages and whom I have referenced in my previous work, *Unveiling the Green*. Charubel (1826–1908) inspired me; his methodology has worked well and helped me to progress in my path. His seminal book, *Grimoire Sympatheia* (2nd edn, 1906), gives instructions and methods on how to contact the spirit of a plant. The book is overtly Christian in its outlook and is subsequently written in the language of the time. I have therefore included passages from it concerning Charubel's methodology, which will allow you to see the similarities and differences between my own experiences and those of Charubel. I have also supplemented my work with the astrological notations from Culpeper, and whose knowledge of astrological and medicinal correspondences is second to none. I also include a botanical drawing of the tree and a meditational image created from my interaction with the tree spirit itself.

Plant sigils are not just marks on paper; they are doorways. Each line, each curve, is a key that unlocks the quiet, hidden spirit of the plant it represents. When you work with a plant sigil, you are not simply staring at a symbol; you are stepping through a threshold, inviting the plant to meet you in a space that is both psychological and psychic. It is a conversation, a shared breath, a weaving together of your spirit and theirs.

This work is not about forcing the plant to speak in human terms; it is about opening yourself to how the plant chooses to communicate with you. And this is the beauty of it—no two people will ever experience it in quite the same way. For some, the plant will come as a rush of scent, sharp and green, or as the ghost of its taste lingering on the tongue. Most often, it is feelings—deep impressions, pulses of knowing—that rise through you, which the mind then translates into something we can understand as words or images. But that translation is only the surface; underneath, the plant is speaking in its own ancient language, and you are learning to listen.

Personally, I use Charubel's sigils whenever one is available, as his work carries a particular depth and resonance that I find invaluable. However, I also create my own sigils for personal use when no Charubel sigil exists or when the plant wishes to communicate with me in a way that feels unique to our relationship. In the plates I have provided, I have included only Charubel's sigils where he has worked with the trees, and where he has not, I have added my own. Please feel free to use them as they are, or better still, develop your own as you work with the trees. This is, after all, a living dialogue, and your personal connection with the plants is what will bring the sigils fully to life.

A brief note on faces

As a coven, one of the things that we noticed when working with a particular plant is that both the 'Name' and the 'Sigil' may be very different to each person involved.

So, "doesn't this invalidate the result?" I hear you say … Let me explain in more detail.

Hi, I am Sian.

To my Husband, I am Siany; my face to him is loving and giving (unless he has done something stupid, then it's homicidal and psychotic). To my daughter, Cerys, I am Mam, cuddly, and a bit scary when she has done wrong, but I love her unconditionally. To my Coven, I am a leader.

All this means is that I have many faces, and the face you know me by greatly determines your view of me and how you interact with me. This is precisely the same for the plants. You also remember that plants are living beings, and if you imagine meeting tree spirits at a party, you would not get along with some of them. Alternatively, you might meet someone when they are ill, and the interaction would be less than satisfactory.

Remembering that you are not the most important person in the relationship is essential, and that some plants just will not want to work with you, although it may be a little ego-bruising, is something we must accept.

And so, with this in mind, I will now share and give an example of my experience with the common buttercup as taken from my previous book *Unveiling the Green*.

Personal experience

I went into the meditational work with the buttercup before reading any of the commentaries on the plant. I thought it was an herb of the sun, all yellow and happy-looking. Well, that was a BIG mistake. Working with this plant was probably one of the most revealing works I have ever done. Buttercup is not a herb of the sun, to start with; it is Martial, beyond a doubt. It stings your hands with its juices, its stem is rough and jagged, and its personality is so aggressive that it's challenging to describe.

Meditational experience

He came to me as a soldier, a corporal, a low-ranking member of the army hierarchy. He reminded me of that terrier you find on farms, which thinks it can rip off your leg before you can get onto the land. He was aggressively possessive about the land. He did not want to work with me and would neither give me his name nor his sigil.

I am describing my experience with him so you can see that sometimes plants just want to be left alone. Charubel described him as attracting the evil mood of psychopaths and those in depression and melancholy. I can see that the feeling I got from him was that he genuinely believed he could take on the world. I feel that if the nuclear holocaust comes, buttercup and cockroaches will be left ruling the land.

> *Ro-Va-Mal*
> *Fuck off, he screamed*
> *Brandishing bayonet and hat equally aggressively*
> *Fuck you and all who come after*
> *Tiny soldier of Mars*
> *Screaming in the wind*
> *Daring the rain to fall*
> *Happy Face hiding Fangs and big balls*
> *Come on you Bastards, Come on!*[44]

It is therefore essential to acknowledge that, although I do not have a working relationship with this person, the contact was there; however, he simply did not want to work with me. Thus, we don't always get the results that we think we otherwise might deserve. With this in mind, I invite you to the section of the book that introduces the trees and shares my personal experiences with each of them.

---

[44] Sian Sibley, *Unveiling the Green: Working Astrologically, Alchemically and Psychologically with Plants* (Black Lodge Publishing, 2022).

CHAPTER SIX

The Oak

DUW-ÁRCHUA

Oak

The oak in magical folklore

One of the most famous oak tree traditions is in that of the creation of *Blodeuwedd*. As the 'flower-faced one' her symbol is that of an owl. Her magic stems from the relationship between Llew and his mother Arianrhod.

Blodeuwedd is a character from the Mabinogi tales. The *Mabinogi* is a collection of Welsh tales divided into four parts, also known as 'branches'. Her story is told in the fourth branch of the *Mabinogi*, '*Math fab Mathonwy*' (Math, son of Mathonwy). According to legend, a woman named Arianrhod placed a *'tynged'* (fate) on her son, *Lleu Llaw Gyffes*, which meant that *Lleu* could never marry a human woman. However, *Lleu's* uncle, *Gwydion*, a renowned wizard, created a wife for him out of flowers and named her *Blodeuwedd*. She was created from the flowers of the oak, the flowers of the broom, and the flowers of the meadowsweet.

Blodeuwedd and *Lleu* got married, but trouble was soon on the horizon. *Blodeuwedd* had fallen in love with *Gronw Pebr*, a warrior and hunter from Penllyn. *Blodeuwedd* and *Gronw* plotted to kill *Lleu*, but this was eventually discovered. *Lleu* was restored to life and took revenge on *Gronw*. As punishment, *Blodeuwedd* was turned into an owl, which is still known as *'Blodeuwedd'* in Wales.

The feminist interpretation sees *Blodeuwedd* as a symbol of female empowerment. She is a woman created by men to serve a man, but she rebels against this role and asserts her own agency by falling in love with someone else and goes on to plot and kill her husband. This tale resonates with me as being similar to that of Lilith and Eve in the Garden of Eden, where both Lilith and Eve are created as companions to Adam, and each rebels in their own way. Lilith screams the secret name of god and flies away, while Eve will receive the serpent's knowledge of 'good and evil'. In both traditions, trees play a central role.

It is therefore not surprising that the oak tree and trees in general, play decisive roles in ancient traditions. Revered across Europe, the oak was sacred to supreme deities such as Zeus, Jupiter, Dagda, Perun, Thor, and Taranis, gods associated with thunder, rain, and lightning. Their association with the oak is no accident: oak trees stand tall, hold moisture high, and draw down the skies, earning them frequent strikes by lightning.

In ancient Celtic lore, the oak was a portal to divine power. Pliny the Elder recounts the story of white-robed priests ascending the oak to harvest mistletoe, believed to have been planted by lightning, and crafting elixirs for healing and fertility. Oak groves became the theatres of ancient magic, where the earth whispered back to those who listened.

Over time, oak crowns became symbols of victory: Roman generals and medieval kings donned them after triumphs, claiming the oak's power and authority. Folklore even offers weather prognostication in its leaves:

"If the oak before the ash, then we'll only have a splash; If the ash before the oak, then we'll surely have a soak."

Culpeper's commentary on the oak

It is so well known (the timber thereof being the glory and safety of this nation by sea) that it needs no description.

<u>*Government and virtues*</u>. *Jupiter owns the tree. The leaves and bark of the Oak, and the acorn cups, do bind and dry very much. The inner bark of the tree, and the thin skin that covers the acorn, are most used to stay the spitting of blood, and the bloody-flux. The decoction of that bark, and the powder of the cups, do stay vomiting, spitting of blood, bleeding at the mouth, or other fluxes of blood, in men or women; lasks also, and the nocturnal involuntary flux of men. The acorn in powder taken in wine, provokes urine, and resists the poison of venomous creatures. The decoction of acorns and the bark made in milk and taken, resists the force of poisonous herbs and medicines, as also the virulency of cantharides, when one by eating them hath his bladder exulcerated, and voids bloody urine. Hippocrates saith, he used the fumes of Oak leaves to women that were troubled with the strangling of the mother; and Galen applied them, being bruised, to cure green wounds. The distilled water of the Oaken bud, before they break out into leaves is good to be used either inwardly or outwardly, to assuage inflammations, and to stop all manner of fluxes in man or woman. The same is singularly good in pestilential and hot burning fevers; for it resists the force of the infection and allays the heat. It cools the heat of the liver, breaking the stone in the kidneys, and stays women's courses. The decoction of the leaves works the same effects. The water that is found in the hollow places of old Oaks, is very effectual against any foul or spreading scabs. The distilled water (or concoction, which is better) of the leaves, is one of the best remedies that I know of for the whites in women.*

The name for Oak in Welsh is Dderwen, and in Gaelic it is Duir. Medieval surgeons used to use Oak galls and other chemicals to clean gangrenous wounds. Oak is still used today by medical herbalists to treat severe diarrhoea and for reducing inflammation.

Charubel on the oak

The Oak (Quercus robur) tree is of every nation, every climate and of every age. The tree of renown; I shall do it service by only using material of authentic renown.

To accomplish this object, I shall give the radical name of this tree by which it was known at the infancy of history and as described in the most venerable records of a hoary antiquity.

In the Hebrew Bible this tree is designated Ashel. The root of this name is supposed to be Gashel. The Hebreu' letter Ash, or AS, conveys the idea of light, or fire; not ordinary iight or ordinary fire; but, that primal light, and first of fires, that ever burned, before the light of day, or dawn of time. The letter EL, or AL, ever alludes to those primal powers, or celestiai intelligences, whose residence is the entire: that region in which revolve the sun, and moon, and all the host of heaven. The EL stands for what, in the Hebrew tonglre, signifies the gods, thus when the two letters are joined—Asrr-EL, whose literary translation into English would be the god-fire, or god-light, as the letter Ash would bear earlier the light signification or the fire interpretation; but the latter appears the more correct.

Such is the radical idea that the name of this tree conveys in the Hebrew language. There is another idea which applies to the Oak, but in the meantime, is to be taken only in a secondary sense, that is Defender or Mediator. We are consequently naturally led to conclude with the question that finding the Hebrew language attaches such an idea to the Oak; are there any authentic records of the Hebrew nation giving to the Oak divine honours? Yes, verily, we have such a record in no less an authority than the Hebrew Bible to shall point out, by way of reference, a few 76 passages having a direct bearing on this subject.

In the 21st chapter of Genesis, it is thus recorded: "And Abraham planted a grove in Beer-Sheba, and called there on the Lord, the everlasting God." This grove consisted of Oak exclusively and was not a mixed class of trees on the fashion of the modern parks which are to please the eye. This grove could not have been for the luxury of salubrious shade, as the *young plants could not have attained such dimensions in the lifetime of the patriarch.*

In the 31st chapter of the 1st Samuel we have an account of the bones of Saul and his sons being buried under an Oak, in Jabesh-Gilead, after being cremated. But the Bible student will be familiar with other passages of a like order, giving unmistakable proofs of the Oak being regarded as a sacred tree in the earliest ages or times of Hebrew history. Additionally, to the sacred records, I will refer you to what is designated, by way of distinction, profane history. Here we have an account of Romulus opening a famous Asylum between two groves of Oak.

In the early days of Greece, we learn of the famous oracle of Jupiter, at the Oaks of Dodona. Among both Greeks and Romans we find the Oak set apart as sacred to Jupiter. But it was in Britain where Oak was preeminently adored in the far past, by our forefathers, the ancient Britons, as related in their feasts and festivals. More especially the grand annual festival which answers to our Christmas-tide, and from which our Christmas has been borrowed.

As Jupiter, who owns this tree, is said to be the father of the gods, even so is the Oak, the king among trees. The Oak being the tree of Jupiter could not be otherwise than sacred in the estimation of the devotional Briton. Nor is that reverence yet extinct; no, it is not altogether a thing belonging to a mythic past, as, to my knowledge, there are a few in this country, at this day, who pay more than ordinary reverence to this tree of millenniums!

We are led to suppose, by the Ministers and Priests of Modern Christianity that the reverence which the Druids, among the Britons paid the Oak, was idolatry. If they were idolaters then are the Ministers and Priests of a meaningless ritual idolaters. Why that reverence for a building made with hands? Why these attitudinal changes, or genuflections, in the presence of decorative art, or ephemeral ornamentation?

If these deserve being called sacred; how much more sacred those sombre shades, afforded by those ample bows of this father and king of the forest trees.

The Oak, as already stated, is the tree of Jupiter. Jupiter as estimated by Astrologers is the largest and greatest benefit of all the plants in the Solar System. In the meantime, Jupiter, as regarded by sages of Prehistoric times, Jupiter was the King of Space, and Director of the Forces of Nature.

Jupiter was the representative of the one true God in the classic history of the Greeks and the early Roman.

Thus we find the one grand festival of the Druids was celebrated at the time of the Winter Solstice; that is when the sun, by apparent motion enters Capricorn. When the sap of trees begin to ascend. It is here we find the Christian Fathers, in order to reconcile the supposed heathen to the new religion, turned this festival to be the commemoration of another sun; the "Sun of Righteousness."

Everything belonging to the Oak was held sacred. Hence, the Mistletoe that grew on the Oak was held in the highest veneration by the Druids on the day the sun entered Capricornus, or rather when it came to the point designated as the Winter Solstice, when a grand procession was formed. The Druid priest, in his white robes, went into the forest in search of the Divine Mistletoe that grew on the oak. When this was discovered, a white sheet was spread beneath the tree. The priest mounting the tree, and with a golden sickle cut off the mystic "Branch," and let it fall into the linen sheet. General rejoicing now began. A young bull was tied and offered a sacrifice to the newborn sun. This was the festival of the Druids, which subsequently was termed Christmastide.

This festival had been celebrated thousands of years before the birth of Jesus the Christ. I would here ask the question; is there at this day any special psychological power in the Oak? It lies within the limits of my privileges to answer this question in the affirmative.

Yes, this is the Psychic cause of Briton's ascendency over the nations of the earth in the past. Britishers! You are not alive to this occult fact.

The Oak is the living Talisman which accounts for the superiority of this country. Please note the following: At the time when our Oaks were abundantly scattered over this land, Britain, as a nation, stood alone. Not only as Mistress of the Seas, but Master of the world!

Flatter not yourselves with the fabricated chimaera that England's wealth makes her greater today than she was centuries ago. Verily as England has increased in money she has decreased in veritable manhood. Why is this so? Answer: her country is despoiled of her Oaks. The father of trees is cut down, and their ancient site has not been replenished with young ones. Since those days when avarice usurped the domain of veneration; when England began to exterminate her spacious Oak forests; from those days date her decline.

Think of these remarks as you may; look at them in whatever light you choose. Call all I have written but a tissue of superstition. The facts are there all the same. I find one satisfaction amid our iconoclastic race: in the parks of our old nobility the grand old Oak may be seen in his primal greatness.

<u>The Healing Power of the Oak</u> *The strengthening influence of what this tree is capable of affording. The following complaints are among those for which it is specially applicable: prostration after long illness; a sensation of an all-gone feeling; a giving up of all, and everything, great timidity; and a constant dread of death.*

For each and all these complaints the Oak is the remedy. The aura of this tree is deep golden. It very much resembles the aura which belongs to the sun; hence the applicability of its primal name, the fire of the gods. The solar influence of the Oak transmitted directly to a mortal in the body would be too positive, and would bring to bear on the organic structure more life force than that organism could endure, which implies sudden death to the subject of that influence.

It thus happens that the cause of sudden death is not for lack of life force, but the lack of an instrument capable of holding that force. It is therefore, in most instances, that more attention should be paid to repairing deranged or damaged tissues, and the removal of obstructives, than to the augmentation of the life forces. Why is it that sudden death is very often the sequel to a state of excitement? it is—in part at least—when the bodily organism is excited there is extra friction, friction beyond that ordinary friction which nature requires, and, as the ordinary life currents are kept in motion by friction, so that when more than the amount of ordinary friction is produced, a more than ordinary life current is stimulated into activity. Thus the strain becomes too great for the organs to endure and sudden death is the result. The Oak is a medium for a special solar ray, which, mingled with order elemental essences, is capable of imparting to the man who may have the requisite wisdom, new life and renewed vigour. The invocative word belonging to the ritual for the Oak is DUIr-ARCHUA, which should be repeated seven times, slowly and reverentially, with face to the North.[45]

Magical methods and use of the oak

Oak is an excellent magical ally for protection spells, but he can also remove negativity from your energetic space. Making small twig bags using fallen twigs and placing them in the four corners of your property is a great way to prevent anyone with ill intent from entering your house. Additionally, you can perform a protection ritual and create a talisman using the protective abilities of the oaks in your area.

---

[45] Cf. Charubel, *Grimoire Sympathia: The Workshop of the Infinite*, 2nd edn (original entitled *Psychology of Botany, Minerals and Precious Stones* (London: I-H-O Books Ltd, 1906).

Ritual protection using oak

Invoke your deity as you see fit.

Now convert letters to numbers on the Jupiter Kamea or Planetary square.

Invocation to the spirit of the oak

> Great green oak, Dderwyn, Duir, Drys
> In the name of Jupiter, who gives us joy
> I ask you to protect my house
> My family and all the spirits that live within
> By your strength, defend us
> By your love, protect us
> I exchange my service for yours
> In light and in the name of the green
> I ask this of you.

After the invocation, write your request on paper and remove all the letters that occur twice.

P~~r~~o~~tec~~t my ~~h~~ouse ~~a~~n~~d~~ ~~all~~ who d~~w~~ell wi~~th~~in.

Then use an alphabet to numbers conversion (to make it easier I have included one here):

A	B	C	D	E	F	G	H	I	J
1	2	3	4	5	6	7	8	9	10
K	L	M	N	O	P	Q	R	S	T
11	12	13	14	15	16	17	18	19	20
U	V	W	X	Y	Z				
21	22	23	24	25	26				

Prmynoewh = 16, 18, 13, 25, 14, 15, 5, 23, 8

Reduce all numbers over 15, by adding the digits together, for example 16 = 1 + 6 = 7

This reduces to
7.9.13.7.14.15.5.5.8

Then plot the numbers onto the Jupiter Kamea using straight lines.

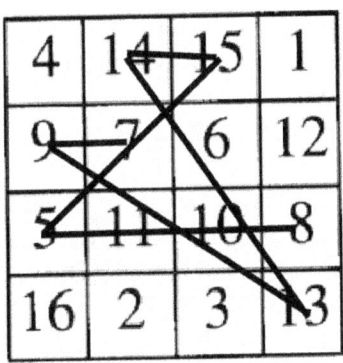

The Kamea to be used in this case is that of Jupiter, a 4 by 4 magic square—therefore numbered 1 to 16 (the traditional squares used for planetary energy magic are 3 for Saturn, 4 Jupiter, 5 Mars, 6 Sun, 7 Venus, 8 Mercury and 9 for the Moon).

When completed, you can refine the design to make it more artistically pleasing. This will allow you to concentrate on it and then activate it later. Now, place it on to either a piece of paper, or draw or even sew it on to a piece of material and then use it as a protection. You can also use it around the boundary of your home or place of magical *practice*.

My own magical relationship with the oak

As you can imagine, the oak is a significant tree to me. My coven is called *DragonOak*. To me the oak tree is a protective spirit who watches over all 'walking the craft path'. He is a male spirit that gives off protective energy to ward off those who would harm us. In my experience with the nature of the oak, it is an intense plant with several levels of involvement, depending on how one approaches and activates its immense power.

Astrologically, oak falls under the planet Jupiter. Still, for me, the oak is Jupiter **in** Capricorn, a guardian who understands limits, boundaries, and the power of structure. The oak doesn't fling its gifts wildly; it holds them, contains them, and uses them wisely. It is protective, a tree that holds back threats and encloses what it loves.

You can see this even in its medicine. Oak bark, with its astringent bite, pulls things inward—it closes wounds, stops bleeding, steadies the body by holding back excess. It does the same with water, tightening the kidneys, stopping what spills away too freely. This is Jupiter's protection but tempered with Capricorn's restraint.

I feel that same protective strength when I work with oak for my coven or for myself. Oak pushes me, sometimes gently, sometimes firmly, to examine what I truly value, to take care of what matters in the natural world, and to offer my service where it is needed. Oak is not about endless expansion; it is about right action.

The oak is a great ally when you are seeking to encourage goodness and wellbeing, but it demands honesty. It doesn't let you hide behind excuses. It asks: *Who are you? What can you actually do to help?* That is oak magic in its truest form—a call to service.

Oak also teaches that magic is not always a solitary path. Its roots remind us that the forest thrives together. You can see it in the soil: the silver threads weave root to root, tree to tree, strengthening the whole. The oak shows me that when the group thrives, the individual also thrives. But if you hoard your magic, if you act only for yourself, the web weakens, and the magic fails.

This is the lesson oak gives me every time I sit with it: be strong, be honest, be of service. Magic, like the forest, is a living system. We either tend it together, or it withers.

A personal meditation with oak

I sit beneath him. Beetles and spiders carry on their business on the ground, busy in their own day. He stands with a broad base, and his rough bark rubs against my back.

As I move my own energy to meet his, I descend into a feeling of calmness, he holds me secure, telling me that no one can see, hear, or hurt me in any way while I am here within his company. His roots around me dive deeply into the earth, where the earth provides all we need to survive, thrive, and grow. As I understand this, I am grateful for her bounty.

The oak now appears as an old man, pipe in hand and stick in his fist. He sits beside me, and I can smell his pipe smoke. He is comforting and reminds me of my grandfather Bryn. My meditations are in the spring, so he smells of growth and the scent of the soil when it first awakens, enlivens the drive towards life.

He does not communicate in the same way as plants do; he conveys feelings and expresses himself through metaphor. I see castles falling, the earth's laboured breathing, the millions of people trapped in cages of their own making, who do not understand that the Green is home, trapped within their own imagined hurts, their inner discourse telling them that they are not enough and how we need more because our neighbour has the new thing and we need it. Don't we?

We must look past the illusions we have placed upon ourselves and see our role. We are shapers and doers, protectors of the earth, capable of creating and moving things. Our imagination is both our greatest ally and our most able foe. In imagining our gods, we created monsters in whose name anything is allowable. Instead, look to the Earth, our true mother, the womb from which all life comes. What we imagine comes to be; that is our gift to the world. We create with thought. But we need to know that our place is secure; we do not need to repeatedly prove ourselves. We need no power over others; control over our actions is enough.

Once we can see our place, we can start to work and open the door to new perceptions. See the magic of the land's life force, and work within it to add our own.

Imagination in Service to the Earth

Oak

The Old Green Man of the wood
Stretching out long fingers to the sun
Holding strong
Protecting the green, raising spirits
Long-lived, lightning-blessed Green spirit of wildness
Deep-rooted Joy of the land
Druid friend and Dragon's door
Give us strength in all endeavours
Protecting those loved and those unloved, both
Open the door in my mind.
Allow me to create wonders.
Dreaming life into reality
Shaping the earth with my thoughts
Moving my thoughts from selfish desire to the creation of beauty.
Show us the way to serve the Mother of all things
The green-girded goddess of all.

CHAPTER SEVEN

The Yew

A
D
O
L
-
R
W
N
G
-
F
A

THE CHURCHYARD
YEW

The yew is a tree of deep mystery, a living link between ancient myths, modern medicine, and the enduring rhythms of life and death. It has always stood as a bridge between worlds—rooted in the soil of the everyday yet whispering of the underworld, of gods, and of the cycles that bind us all. Found in sacred groves and ancient churchyards, the yew has watched humanity for millennia, silent and patient, holding its secrets.

Some of the oldest living witnesses to our history are yews. In Wales, ancient yew circles date back an astonishing 3,000 to 5,000 years. One that I feel especially connected to grows at Llanfeugan, nestled in the Brecon Beacons. It stands within a circle of 13 yews in what is now a graveyard, though it is clear the church is a late intruder—these trees were here long before Christianity laid claim to the site. Walking beneath its dark, twisted branches, you can feel the weight of centuries, as though the very air hums with memory. It is not just a tree; it is a bridge—between life and death, between the human world and something far older, deeper, and wilder.

The yew's very nature embodies this liminality. Evergreen foliage speaks of immortality and eternal return, while its poisonous qualities root it firmly in the shadowed realm of death and transformation. In mythology, it is almost always a tree of thresholds, marking the passage from one state of being to another. The Druids regarded yews as sacred, seeing them as protectors and guides through the veil between worlds, a source of wisdom that could reach from the land of the living into the mysteries of the beyond.

This ties so beautifully into the theory that Yggdrasil—the Norse World Tree—may have been a yew rather than an ash. Some early descriptions refer to Yggdrasil as a *needle ash,* a curious term since no such tree exists, but which fits perfectly with the yew, with its dark, needle-like leaves. For me, after years of working closely with this ancient being, this theory feels absolutely right. The yew is a tree of life, death, and rebirth all at once; it holds within it the whole kosmos, just as Yggdrasil was said to.

Even yew's physical form speaks of this eternal cycle. When its branches touch the ground, they root and grow anew, cloning themselves, whispering of immortality. One of the most beautiful myths that captures this idea tells how, at Adam's death, three yew arils were placed in his mouth. From these grew three sacred trees: the White, the Green (or Golden), and the Red. The White tree became linked to the Staff of Moses, the Green to the burning bush, and the Red to all yews living today. These stories, along with tales of gods born from trees, are explored in Janis Fry's *The God Tree*—a rare book now, but one I would strongly recommend seeking out if you can.

The yew as healer and poison

The yew truly embodies its mythological reputation as a tree of both life and death. It harbours poison yet gives life-saving medicine. Traditionally used in Chinese medicine, the Pacific Yew (*Taxus brevifolia*) revealed a compound that modern science named *paclitaxel*—better known as Taxol. Taxol works by disrupting the cellular mechanism of life at the cellular level: it binds to microtubules, the tiny structures that allow cells to divide, thereby freezing them in place so that cancer cells cannot complete mitosis. This ultimately forces them into apoptosis, or programmed cell death. Breast and ovarian cancers, which divide rapidly, are particularly susceptible.

Another compound, Tamoxifen—used to treat several types of cancer, particularly breast cancer—also originates from the yew. It is a plant that can take you to the edge of death and bring you back again. But let me be clear: the yew is far too poisonous to be used as a herbal remedy, and I would never advocate taking it in any form. Its gifts are not for casual use; they are sacred, hard-won, and often come with a price.

The yew in magic and necromancy

The yew has long been regarded as one of the most potent magical trees, holding a unique position as both a guardian of the living and a guide for the dead. Its very nature speaks of thresholds: evergreen yet poisonous, ever-renewing yet ancient, a tree that embodies the endless cycle of life, death, and rebirth.

Traditionally, the yew was sacred to chthonic and liminal deities, most notably Hekate, the queen of witches and mistress of necromancy. Its Saturnian correspondences make it a powerful ally in bindings, both physical and spiritual. It has been used to 'hold' magically—sealing contracts, curses, or protective wards—but also in unbinding, particularly to release souls from magical or spiritual bondage. Through ritual acts, practitioners have used the yew to elevate and free the spirit of the dead, a process that reflects its role as a tree of passage rather than imprisonment.

The association of yew with necromantic practice runs deep. Ancient Druids and later cunning folk carved runes and ogham from its wood, believing it amplified divinatory and spirit-communication work. In necromantic rites, yew was burned or used as incense to raise the dead, summon ancestors, or calm restless spirits. The Pacific Yew was particularly valued in Chinese and shamanic traditions, where its smoke was considered a gateway to the underworld.

Planted in churchyards and graveyards across Britain and Europe, the yew acted as both protector and guide. It was believed to ward the living from errant spirits and to draw malign forces away from the dead, keeping graveyards spiritually 'clean'. This is why yew often stands sentinel at the south and west walls of old churchyards, places traditionally associated with the dead.

Astrally, yew has long been used for spirit travel. Wands, staffs, and ogham staves carved from its wood are employed in journeying rites, opening the practitioner to visions of the underworld or enabling communion with spirits. Some traditions speak of yew as a "World Tree" in miniature, each staff acting as a bridge between the planes of existence.

Necromantic plant-magic, or plant-based death work, places yew among the most powerful of the baneful allies. The tree is considered to live and die at the same time, existing in a state of liminal balance. For this reason, yew is sometimes described as a 'master of life, death, and rebirth', capable of revealing the mysteries of cyclical existence to those who dare to work with it.

In all such practices, caution is essential. The yew's baneful nature is not symbolic, it is highly toxic. Its spiritual potency mirrors its physical danger, and all magical work involving yew should be approached with respect, care, and a deep sense of reverence for its role as a gatekeeper to the Otherworld.

In view of its association with Saturn, you can use it to bind and hold magically. As noted, it can also be used as a tree to affect magical bindings, and particularly in unbinding, which results in the freeing of someone from bondage or enslavement, achieved by raising that person's soul through a magical act.

A binding spell to stop malfeasance

There is considerable discussion about the ethics of binding someone who is causing harm. I do not believe that the *'Harm None'* ethic is viable in a magical sense. You cannot harm anyone; it is both physically and spiritually impossible. If I disagree with someone's spiritual perspective, e.g. an evangelical Christian, I am hurting them by challenging their perception of the omnipotence of their god. I cannot physically harm anyone as I eat and have a presence in the world. By doing that, I am harming the plants and animals I eat, and the bacteria I kill by consuming antibiotics *ad infinitum*.

Harm none is not achievable. However, harm as little as you can and when you do need to harm, take full responsibility for your actions; for me, this is a viable alternative and agrees with my personalised moral compass. So, when working on binding with yew, it is essential to understand that the yew is a tree that 'raises the soul'; it frees us and when we ask it to work on a project where a person is released from abuse and violence, it is an act that can elevate that person and give them freedom—this is indeed acceptable for the tree.

However, using it to bind for another reason, for example, something that did not advance the spell recipient and might constrict their freedom, would not be advisable.

The following spell is one that is used to help a woman whose husband was abusive both physically and emotionally.

One needs an object link for this spell. Thus, an object link should be something belonging to the man and where possible directly from the man, such as something with his fluids or skin on it including hair, nails or the like. Wrap or enclose the object (or hair around) a piece of kidney, liver or heart. Place the meat into a box made of yew as a limiter or 'boundary' between the man and the woman. Place mirrors on the top and sides of the box. The woman in this instance, then buries the box on their property.

The spell

I call upon the Powers Dark
To cast into the night so black
A binding strong and fast and tight
To bind your harm and stop your might
To give your woman calm and peace
To hold your tongue, say not your piece.
For if you harm her one more time
Into my maw, your soul will climb

and hold you fast with my tight grip
And from your body soul shall rip

Your tongue and trammel harm they do
And so return the harm to you
Into this meat, your dirt and grime
I put and so decay by time
And as the meat shall rot away
So does your heart and soul now pay the harm you do to kith and kin
Now does that harm come back within

By guardian of the darkest night
And owl that screams whilst in full flight
By Ywen's strength, I bind you tight in Chains forged from your own sheer spite.

And so now, man with conscience small
You are bound up by one and all
Your words and violence hold in thrall
I bind you to all harm you do and so that harm comes back to you
Time on time and act of act
I bind you now; it is a fact
As I will so mote it be
Now it is your choice what end there be.

(NB the wording of this work: this magical work will not activate until there is a violent act in speech or physicality on the woman. If he does not behave violently, the spell will not take hold. It is entirely a work of his own making.)

Culpeper's description

It grows to be an irregular tree, spreading widely into branches. The leaves are long, narrow, and placed with a beautiful regularity. The flowers are yellowish, and the berries are surrounded with a sweet juicy matter.

<u>Place:</u> *We have it growing in woods, and in the gardens, but its usual ancient residence is the churchyard: conjectures upon the antiquity and origin of which plant, has brought forth much pedantic nonsense; Gray observes this in* The Grave, a Poem, *when he says,*

> Well do I know thee by thy trusty Yew,
> Shading for years thy gloomy church-yard view;
> Cheerless, unsocial plant, that loves to dwell
> Where scatter'd bones man's dissolution tell.

<u>Government and virtues</u>. This is a tree of Saturn. The leaves are said to be poisonous; but the wood, if it grew with more regularity would be very valuable. This tree, though it has no place among the physical plants, yet does it not deserve (at least in our climate) so bad a character as the ancients give it, viz. a most poisonous vegetable, the berries of which threaten present death to man and beast that eat them; many in this country having eaten them and survived. However that be, it has very powerful poisonous qualities, that rise by distillation. In this form it is the most active vegetable poison known in the whole world, for in a very small dose it instantly induces death without any previous disorder; and its deleterious power seems to act entirely upon the nervous system, and without exciting the least inflammation in the part to which it more immediately enters. It totally differs from opium and all other sleepy poisons, for it does not bring on the lethargic symptoms, but more effectually penetrates and destroys the vital functions, without immediately affecting the animal. These observations would not have been made, or the article inserted here, but to caution against any rash application of it, for though it is sometimes given useful in obstructions of the liver and bilious complaints, those experiments seem too few to recommend it to be used without the greatest caution. The deleterious qualities of laurel-water are more than equalled by this.

Charubel's commentary on the yew

The pedigree of this tree and that of its numerous allies is a parallel with the Fir tree, to which it bears a rather striking resemblance; but which is, after all, of a different order.

The Yew is an order of plant life that partakes of both the Fern and the Fir. It may be said to form a connecting link between these two. In the order of cosmic development the Yew stands anterior to the Fir or the Pine. The Yew adapts itself to any country, and that without but very few changes in its structural appearance.

The Yew grows to a large tree in China. In Japan its leaves resemble the foliage of the Maiden-hair Fern; in the meantime it is with our *grand old English Yew that I am now concerned. This tree grew in Britain ever since it was an island. It may be truly said to be indigenous to this island, and grew here when no other save the Fern tree abounded.*

Thus, the Yew was the first of the flowering and fruit-bearing plants then on this island. It thus stood forth as an index finger pointing to a new state of development, while its leaves point, to some extent, to a dead past, its fruit points towards a better and more hopeful future. While the leaves of this tree are poisonous, the fruit is luscious, and are eaten by birds and children. This tree absorbs and dispenses. It absorbs the death principles of its surroundings, and gives out the life principle. Thus you may perceive, in what I say of this tree, that its being the connecting link between the Fern and the Fir, it is rational to suppose the one nature should absorb the evil, and the other should give out the good. The Ancients, in the far past, looked on this tree as being the Symbol of the planet Saturn and the Sun. Saturn stands for all things mortal, or the termination of the earthy. The Sun is the beginning of a life that shall not be subject to decay and death. Thus the English Yew as it stands at this day in our old graveyards; more especially the old churchyards of Wales, where it abounds to a greater extent than in England; is a very proper Symbol for the mortal and the immortal; death and life; it has been planted by pious hands in our rural village graveyards, as a fit and proper symbol of life and immortality which came to light by the Gospel; and has been made use of in the past ages as the emblem of the Messiah, who took on Himself that nature which had become the subject of death, and in the meantime, by those higher powers of life which He possessed, developed the immortal. Thus out of death came life, as it was out of darkness that light came. There is a very pleasing disparity between the beautiful pink berries of the Yew tree and its sombre evergreen and poisonous leaves. There is a much more pleasing disparity between the mortal and the immortal in man. The mortal descends, but the immortal ascends. No one can tell the age of some Yews which I have seen in some of those little out-of-the-way churchyards among the mountains of Wales. I counted seven in a small churchyard among the hills in Denbighshire. But these grand old trees abound and occupy large spaces in most country churchyards; they give a very picturesque aspect to rural villages, which I consider calculated to produce a very pleasing effect on the soul of the thoughtful.

The Psychic Virtues of the Yew Tree

On the psychic plane, this tree appears in a different costume. The dark-green foliage is transformed into golden, mingled with azure.

From every branch, I see a small hair-like stream descending of what looks like Crystalline Nectar. I see an innumerable host of fairy-like beings about this tree, resembling a species of diminutive humanity. Each of these tiny beings is drinking at these crystal currents.

These beings appear to be primarily allied to the Yew tree. They are not only feeders on this tree themselves, but they can be made the distributors of its virtues to those whose soul may have an affinity to the sphere of the Yew. I do feel grateful to heaven for so glorious a revelation; and although but few of my readers will be sufficiently interested in this revelation to become the recipients of these mystic virtues, yet I, for one, shall be benefited; seeing I make known to all what I receive, if others are nor benefited it will not be my fault. This tree is not especially for any one form of the disease or such as are recognised as a disease. It has more to do with the soul of the individual.

In virtues are expressly to build up the soul, which is the spiritual body. You may have read those words uttered by King David in one of his Psalms: "O spare me that I may recover strength before I go hence and be no more." It is evident that the Psalmist needed his soul strengthened. He must have had, for the time, a glimpse of another interior body, which needed some little repairs before pulling down the old house. There was an epoch in the far past when man lived more on the psychic than on the so-called intellectual plane; this is more than what the present humanity is capable of realising.

Our present conceptions of beauty are not in unison with those possessed by man during one of those buried aeons of the past. Nature closes each door after the birth of her offspring. Thus when one race has completed its round, fulfilled its mission, the door of that degree attained to by that race is closed behind it. The succeeding humanity cannot form any idea as to what may be the conceptions of its predecessor, but the predecessor may form correct notions of its successor. In the meantime, as there are always exceptions to every rule, or in other words, there ever have been those who have lived, while members of the succeeding race, the life of its predecessor, and have been able to realise what may have been the leading ideas of that race. But, when these exceptional characters seek to make these same ideas known to their contemporaries, they are sure to be misunderstood, and what they say or write, for the time, will not be appreciated.

This tree was known to the Psychic race, who were able to imbibe from its branches, as it appeared to them in that life and on that plane of existence, a support which the humanity of this heady race have no conceptions of. Clairvoyance approaches that condition; at the same time, neither Clairvoyance, Clairaudience, nor Psychometry are to be considered identical to that state.

It is vain consulting an ordinary disembodied spirit as to the nature of this Psychic race, much more useless to consult modern scientists. There is a phase of untutored and unspoiled childhood, which bears a more striking resemblance to psychic man than any other outward embodiment that I am acquainted with. Hence the applicability of those words of Jesus: "Except ye become as a little child, ye cannot see the kingdom of heaven." That child to whom the Master alluded was not the spoiled, sharp, or precocious child of this nineteenth century, which I design the heady race, being the antithesis of the soul race.

What I write is but a fragment of the experience of that everyday life I am living on the soul plane. The psychic side of nature with her boundless resources are as familiar to me as the hills and valleys of my native land. I come in contact with more there than can be found here; hence the Yew appears more marvellous to me than it does to you. I am not at all surprised that the good old men of the past should have chosen this tree as Monarch of the graveyard and companion of the dead!

I hope I may never see the day when this heady race will have attained that degree of turpitude as to cut down this Grand Old Relic of a long lost past; and Survivor of the Cosmic wreck. The noble Oak has been partially destroyed; and as a result the present race is weakened on the outer plane. Should any one presumptuously in my presence take up the axe to fell the Yew I will cry out, not sing out, "Goodman spare that tree, touch not a single bough." In it a past eternity unites the present now.

Who among my readers are capable of being benefited by the psychic nature of the Yew? Those who are highly sympathetic; those who are impressionable; those whose minds are not absorbed in the things of the senses, those who are fond of solitude; those who delight in the contemplation of nature where it is most natural; not the most accomplished; not the most highly educated; nor yet the most extraordinary intellect. As a rule, it may be calculated that the more ordinary samples of humanity are suitable for the reception of those influences and virtues that this beautiful tree possesses.

It has been already stated that the Yew tree is a soul strengthener; but this phrase is scarcely sufficiently comprehensive; I will further say, it includes lowness of spirits, or a sense of great depression, and that when there is nothing in the circumstances of the individual to constitute a real cause for such a state. All such symptoms indicate a weak or infirm soul; and when manifested through the brain, or the outer consciousness, is insanity. There is no radical cure for this kind of disorder in the Therapeutics of Medical Science. Nor are there but few of the Profession so bold as to pretend that drugs may cure such. Instead of medicine, they generally advise a change of scenery, cheerful company, or a sea voyage. In the meantime, there must be a remedy, but that remedy must be capable of reaching the afflicted part, or primary seat of such an affliction. I have, in the Yew, discovered a remedy that will absorb that morbific effluvia, which, like the horrid nightmare, clings to the helpless soul—the spiritual body — and at the same time impart to those psychic wounds a healing balm. Connected with the cure are rules and observances to be complied with, as well as an Invocation to be uttered, the whole should be strictly and religiously observed. The best time to apply to the Yew for help is the Seventh hour past noon. The patient or his helper should devote the greater part of the hour to these meditations, either in reading what I have written respecting it or thinking on the tree itself.

Personal observations

Astrologically, I feel that the yew is at its strongest when Saturn is in Aquarius. Here, it teaches you about boundaries, but not as obstacles. Instead, it shows you various ways to address specific limitations. It teaches you how to work with and around them, rather than simply accepting them as they appear.

Yew is a slow tree. It is almost immortal, and this can be seen in the extraordinary length of time it takes one to get to form a relationship of any kind with it—patience is required. But when I did manage to establish a true and beautiful relationship with the yew, the results were terrific. I then learnt about the movement from death into life and back again. The spirit asked me the following: "You all worry about life after death; why does no one ask about life before birth?"

I learned about cycles, when to go out and when to come in, using my intuition. I worked with yew when my mother passed away suddenly, and He helped me with my grief and guilt, and also established the fact that my relationship with my mother had not been the best.

I would also learn that on occasions, life can be challenging and sometimes lead to soul-destroying experiences; yet, the yew would provide a different choice of direction and solution—another door would always be opened, offering an alternative route.

I agree with Charubel when he states that the yew tree is 'a soul strengthener'; for it has personally given me the strength and support during great times of need.

Personal meditation on the yew

He came to me as a crow-headed figure, cloaked in black, leaning on a great staff. The air thickened with his presence, carrying the scent of burnt charcoal and damp earth, the smell of October tipping into November. That rich, wet tang of decay, but threaded through it, the quiet promise of new life waiting in the mulch.

When I sat beneath the yew, the spirit sat with me. It did not speak much, no grand revelations, no riddles—just its weight beside me, steady, supportive, patient. The silence was its teaching. The energy that rolled from the tree was immense, like leaning against a mountain that had been thinking for thousands of years. A being of vast wisdom, of impossible age, and one to whom you offer the deepest respect or you do not approach at all.

The yew teaches time. Not the time we measure, cut into neat little slices with clocks and calendars, but the true time that hums beneath all things. The more we measure it, the more it slips through our fingers. To live is enough. You do not need to prise open the mechanism to understand it. You can dissect a bird and know every tendon, every hollow bone, but you will never feel the joy of taking wing, never hear the wild scream of a kestrel stooping for the hunt, nor the wriggle of the worm rising under dancing feet. To *know* something is not the same as *being with* it.

The hollow of the yew, the dark heart of its trunk, is the doorway to the underworld. It is a gate, a place where the dead walk close, and where the crow-headed man speaks of death in that matter-of-fact way of his. Death, he said without words, is not an ending but a turning. The entropy of livingness is not permanent; energy shifts, circles, moves from one shape to another, one experience to another. Look at the world, he seemed to whisper, through all its eyes—the child's wonder, the parent's care, the elder's patience. There is no end. Only breath after breath, beginning after beginning.

Your last breath here is simply your first breath elsewhere. No fear. No punishment. No hell. Just a step across a threshold into another way of being.

Time is to be lived, not to be measured

Yew

Being of time
Sitting in the land of the dead
Counting the years as minutes by your clock
Scarred and red
The crows find homes within you
And are drawn upon your face
Toads at your base and dragons in your depths
You speak to the soul of all things
Dark are you, wet and deep and slow, so slow
Looking out across the land where Romans ploughed
Yesterday for you, ancient history for us.
We are fleeting in your eye
Buzzing across your sentience
Gone before we know it
While you stand, sentinel to time.

CHAPTER EIGHT

The Ash

HOO-MEL

Ash

The lore of the ash tree

The ash tree has played a significant role throughout the various traditions and cultures in history. It is often associated with strength and protection, and has a deep connection to the spiritual world.

As we noted, in the Norse tradition, the World Tree or Yggdrasil is often interpreted as an ash tree. Yggdrasil is a kosmic tree that connects the different realms of existence, including the human world (*Midgard*), the realm of the gods (*Asgard*), and the realm of the dead (*Helheim*). The tree serves as a central axis of the universe, with its roots and branches connecting the various realms.

In the Celtic tradition, the ash tree is associated with the god *Gwydion* and is often linked to protection, healing, and divination. Celts believed that the ash tree had protective qualities and could ward off negative energies. It was also considered a tree of transformation and rebirth.

In Greek tradition, the *Meliae* were nymphs born from the ash tree after the blood of the castrated Uranus fell to earth.

They are often depicted as protectors of the trees and nature.[46]

In Scandinavian folklore, the ash tree was believed to have protective properties. It was commonly used to make shields and spears, reflecting its association with strength and defence.

The ash tree also held great significance for the Druids of the British Isles. It was one of the *Three Noble Trees* along with the oak and the thorn. The ash was regarded as a protective symbol and was used in protective rituals to ward off negative energies.

In Slavic folklore, the ash tree was sometimes associated with the *Leshy*, a forest spirit or sometimes recognised as a deity. The *Leshy* was believed to inhabit the forests and protect the natural world.

The ash tree therefore possesses both protective and healing properties. Its wood was often used to make amulets, charms, and talismans in order to ward off evil spirits or promote good health.

Ash in healing and magic

The ash has long been a tree of guardianship and restoration, a friend to the fragile and the ill. In many old practices, its very wood was seen as a vessel for drawing sickness out of the body and holding it fast within

---

[46] The *Meliae* or *Oread*-nymphs of the mountain ash tree were born out of *Gaia* (the Earth) when she was impregnated by the blood from the castrated *Ouranos* (lit. Uranus, or more lit. the Sky).

the tree, where the earth could transform it. Sick children were passed through a natural cleft in an ash, and the trunk was then bound up again, sealing the illness away. Once closed, the bond between the child and the tree was complete—the tree took the sickness, and the child walked away healed.

Ash sap, taken carefully, was once given to newborns to protect them from disease—a symbolic blessing as much as a physical one. Ash keys, its winged seeds, were carried as charms to guard against enchantment and fae mischief, and in some instances they were placed in cradles to keep infants safe from being 'swapped' by the fair folk. Even the bark was used magically; warts were burnt with needles and pinned into the wood to pass the affliction on, allowing the tree to bear it instead.

The ash not only protects the body; it also protects the soul. Mariners carried amulets of ash to keep them safe from drowning, and in many places ash was planted near wells and sacred waters to watch over the living and the dead alike. Its presence was thought to calm the restless and guide spirits safely across the threshold.

To work with ash in healing magic is to work with a tree that binds and releases, that holds sickness until it fades, and that wraps its long arms around those who seek its care. But as with all trees of power, respect must be given; ash heals because it chooses to, not because it is commanded. Approach it as you would an elder, not as a means to an end.

In Christian mythology, the first day of Lent is Ash Wednesday. On this special day of reflection, Catholics wear a marking of the cross in ash on their foreheads. The ashes symbolise mortality—"Remember you are dust, and to dust you shall return." But you might be wondering, where do the ashes for Ash Wednesday come from? However, a clear distinction should nonetheless be made in that, the 'ashes' on Ash Wednesday are created by burning 'palms' from the previous year's Christian Palm Sunday celebration. Palm Sunday is the last Sunday of Lent, marking the beginning of Holy Week. It is therefore on this day that people laid palms to cover Jesus's path as he arrived in Jerusalem, days before the crucifixion.

Culpeper on ash

> *Description. This is so well known, that time will be misspent in writing a description of it; and therefore I shall only insist upon the virtues of it.*

Government and virtues. It is governed by the Sun; and the young tender tops, with the leaves, taken inwardly, and some of them outwardly applied, are singularly good against the biting of viper, adder, or any other venomous beast; and the water distilled therefrom being taken a small quantity every morning fasting, is a singular medicine for those that are subject to dropsy, or to abate the greatness of those that are too gross or fat. The decoction of the leaves in white wine helpeth to break the stone, and expel it, and cures the jaundice.

The ashes of the bark of the ash made into lee, and those heads bathed therewith, which are leprous, scabby, or scald, they are thereby cured. The kernels within the husks, commonly called ashen keys, prevail against stitches and pains in the sides, proceeding of wind, and voided away the stone by provoking urine.

I can justly except against of all this, save only the first, viz. That ash-tree tops and leaves are good against the biting's of serpents and vipers. I suppose this had its rise from Gerard or Pliny, both which hold, that there is such an antipathy between an adder and an ash-tree, that if an adder be encompassed round with ash-tree leaves, she will sooner run through the fire than through the leaves: The contrary to which is the truth, as both my eyes are witness. The rest are virtues something likely, only if it be in winter when you cannot get the leaves, you may safely use the bark instead of them. The keys you may easily keep all the year, gathering them when they are ripe.

Charubel on ash

THE ASH TREE (Fraxinus): *This tree is said to be a member of the Olive tribe. Many Virtues are ascribed to this tree by medical practitioners. It is said to be a Ubiquitous medicine for which the powdered bark is used in decoction. It acts upon the Liver beneficially.*

Viewed on the Psychic plain; this tree has a purple, or deep plum colour, which grows darker towards the ends of the branches. The lower part of the stem is a beautiful mauve. Such is the Ash Tree as beheld by me on the soul-plane. There is one very marked peculiarity connected with this phenomenal display: the branches end, in what resembles a very dark fog; by which I learn, that the leaves, and outermost branches, even, in its earthly garb, possess subtle properties; not to be found in the bark, or the lower part of the stem.

It is a well-known fact, as recognized by those living in rural districts, that rain droppings falling from the leaves of the Ash Tree kills all vegetation which may aspire to live beneath its shades. I here approach the main subject, to which every observation is, or should be, subservient, which is, the healing properties on the Psychologic Plane: Acute Pain in the Chest, or affections of the Bronchia—the result of obstruction. Let the sufferer direct his, or her thoughts to the stem of the Ash Tree. But by rendering further assistance to the afflicted one, who may be seeking the remedy; I here give the sigil, or symbol of the Ash, which expresses the tree's mystical powers, properties, or virtues. The word for invocation is HOO-MEL

You imbibe, by faith, the aura, drink in by your inspirational desires this Divine Remedy. This remedy promotes the various secretions that may be obstructing, in this way this remedial agency aids nature, by carrying away various impurities. Beyond all this, the leaves and outermost branches, contains a safe, and sure remedy for Tumours, Abscesses, and all Chronic Excrescences of the Skin.

When such a treatment is required, direct your thoughts, as intently as possible, to the green leaves, in your mind; but as it may not be possible to see an Ash Tree where you may chance be situated, therefore, in the absence of the tree, you may contemplate this symbol, which is the key that unlocks the door, which opens, and admits you into those mysteries which may help you, in the absence of your knowledge of the why's and wherefores. But, additional to all this: the symbol is an expression and is far louder than any vocalised one; for the organ of human speech is incapable of uttering, what this symbol utters. But this word celestial, embodies the forces you need, and, which at the time, places an instrument in your reach, by use of which you may command the needful forces, and they will forthwith come and obey you. But, some one will here reply: What! Am I asked to use a word, whose meaning is hidden from me? My answer to thy question is: If thou rejects all but what thy intellect may be able to grasp, and subsequently hold, and retain; then, I must confess, thy sphere is a narrow one, thy prospects cheerless, and thou art poor indeed; but, if thou insists on an explanation, I have none to give, beyond what has already been given. I need not reiterate my former injunctions, with respect to that reverential mood of mind to be observed in the use of the word here given. You may use it as often as prompted until the benefit you anticipate is realised. Choose the most convenient time for yourself.

In all cases, have the sigil before your eyes, and, what with the contemplation of what you are looking on, conjoined with the word utterance, you will soon realize wonders.

Personal experience of the ash

Ash has a particular place in my heart; it was a tree that lived in my garden as a child, and I often went to talk to it in times of unhappiness and trouble. To me, the ash is the activator in the tree family. For example, Culpeper puts ash under the remit of the Sun, but I would add that it also has 'Martial' qualities: for it increases drive, gives you courage when yours fails, and can drive you onwards when you think you cannot succeed.

For this reason, I would therefore associate ash with the *Sun in Aries*, that first initial push of the Sun as it rises after the coldness and hibernation during the winter.

I have worked hard to protect the trees in my area from the fungus that has decimated the populations across the UK. As a result, my wood, garden and drive are now filled with young ash! You can see how the trees have developed and thrive; they show no sign of the binding and rotting fungus, yet it remains to be seen over the next few years. Although infected, other older trees in the area are still alive and are producing enough leaves to get them at least through the following winter. Owing to the trees' poor health, I do not use ash for magical purposes. Ash needs time to recover, and I am there to provide both magical and physical support by keeping the environment free from things like fly-tipping and ensuring that the ditches around our land are clear, so the environment does not get too wet.

Meditation with the ash

My tree is triple-trunked and, although infected by fungus, still occasionally talks.

The ash I work with comes through as an old woman.

She is grey-haired, and over the last three years, her complexion has improved from a grey-blue to a pinker, healthier colour.

She appears to me in silver and Green robes, looking for all the world like a druid.

She tells me that fighting for the rights of others is essential.

As she says, her life is enhanced by the presence of others, as is my own if I open my eyes to see them. We must leave behind the comfort of solitude and re-enter the community of the earth.

She talks of sacrifice. Her sacrifice of leaves and seeds to feed others.

How she feeds other trees when they are ill and how they, in turn, are feeding her now in this time of illness.

She asks me what I will give up to allow others to thrive, and who I will provide for.

This is a tricky question as I like my 'things'. And giving them up is inconvenient and difficult and opens the door to hard work that, in reality, I am trying to avoid.

She makes me face the truth about what I am doing, sitting and reading books on ecology on my electronic device powered by coal.

She makes me look directly at my practice and face my lies. Those little white ones seem to do no harm but actually make excuses for inaction.

She tells me that if mankind does not start thinking of how we are in the world. That this will impact all others on the planet.

We must sacrifice our gluttony, pride and ideas of superiority for others to survive.

What do we want, and what do we need?

Sacrifice to learn

Sacrifice to live

Sacrifice to allow others to thrive

In sacrifice, we grow

Ash

Standing silver green in twilight
Shining forth with words of truth
Feeding others with your own self
Giving sustenance and youth
Upside down on spreading branches
Perception changed to see it all
A hanged woman sees the damage
To say is not to do at all
The Eyes and heart are all pulled open
watch illusion pulled apart.
And sacrifice it is demanded
In service to the sacred land
Where we live and have our bodies
Where we all must make a stand

CHAPTER NINE

The Rowan

The magical tradition of the rowan

In many cultures, rowan trees are believed to possess protective powers against evil spirits, witches, and malevolent forces. People often plant rowan trees near their homes or wear talismans made from rowan wood to ward off harm.

In the Celtic tradition, rowan trees were considered guardians of the netherworld. They were often planted near ancient stone circles or other sacred sites. The rowan forms a veil, link, or bridge between the mortal world and the realm of the spirits.

Rowan trees are therefore particularly associated with wisdom and vision in Celtic and Norse traditions. Eating rowan berries was thought to grant the gift of foresight, allowing individuals to see into the future or gain insight into hidden truths.

Rowan trees are often linked to fairy lore. For it is believed that fairies use the rowan tree as a gathering point or even as a place to reside. Because of this, cutting down a rowan tree is thought to anger the fairies and bring extremely bad luck on the culprit.

In some cultures, rowan trees are associated with healing properties. Rowan berries are used in traditional medicine to treat various ailments, and the bark is sometimes used to make herbal remedies.

Rowan branches are often used in weddings as protective charms. Couples stand under a rowan arch to receive blessings and protection for their union.

Rowan trees were also linked with navigation and safe travel on the sea. Sailors and travellers often carry rowan wood or berries as amulets to ensure a safe journey and protect against shipwrecks or accidents.

A general description

The Mountain Ash (*Pyrus aucuparia*) is not related to the true ashes, but has derived its name from the similarity of the leaves.[47] Compared to the true ash, it is a small tree, rarely exceeding 30 feet in height. It belongs to the order *Rosaceae* and is distinguished from its immediate relations, the pear, crab apple, white beam and wild service tree,

---

[47] *Pyrus aucuparia* (L.) Gaertn. First published in *Fruct. Sem.* Pl. 2: 45 (1790), where the name is a synonym of *Sorbus aucuparia*, the officially accepted form.

by its regularly pinnate, 'ash-like leaves'. It is generally distributed throughout the country in its natural state, but is also widely cultivated as an ornamental tree.

All parts of the tree are astringent and may be used in tanning and dyeing black. When cut, the Mountain Ash yields poles and hoops for barrels. Both the bark and fruit have medicinal properties. The fruit is rather globose, with teeth at the apex and two to three seeded cells. They are used medicinally in either the fresh or the dried state.

Medicinal action and use

In herbal medicine, a decoction of the bark is used to treat diarrhoea and administered as a vaginal injection in cases of leucorrhoea, among other conditions. The ripe berries furnish an acidulous and astringent gargle for sore throats and inflamed tonsils. For their antiscorbutic properties, the berries have been used in scurvy. The astringent infusion is used as a remedy for haemorrhoids and strangury.

Its fruit is a favourite food of birds. A delicious jelly is made from berries, which are excellent with cold game or wild fowl, and a wholesome perry or cider can also be made from them. In Northern Europe, they are dried for flour and yield a strong spirit when fermented.

Charubel's commentary

The Mountain Ash (Pyrus aucuparia) *is another of those trees which all who have the privilege of living in the country have become acquainted. It is one of the most beautiful trees which adorn the banks of running streams or climes to the giddy heights of mountainous districts. This plant belongs to the Apple family. Its botanical name is* Pyrus auruparia. *It is a plant that will grow at an altitude of nearly three thousand feet above the sea level; at the same time it can adapt itself to the richer soil of the warm and fertile lowlands. Its leaves are what Botanists call pinnate, which means, that the leaves form a shape which much resembles a feather. These pair of serrated (saw-like) leaflets are from four to six inches long. The flower is cream-coloured; the fruit is globular, orange at first, afterwards scarlet. There are three stages in the annual changes of this plant; May and June it is green and loaded with cream coloured bloom which loudly makes known the joy of its youth. After this there are nearly two months during* which period it may be said to have no specific comeliness. But when September dawns on us the fruitage of the Mountain Ash becomes conspicuous; afterwards, that blush of rich vermilion cannot fail to attract the dullest and most indifferent of observers. These bunches of scarlet berries are generally so profuse as to make this tree conspicuous at a considerable distance.

The berries are harmless and full of juice which is intensely acid. All that I know as to the medical properties of this tree is: it was customary with the farmers in my native country when I was a boy to press out the juice of the berries and bottle it, to give calves for a complaint which the people called the gurr; I am inclined to consider this word taken from gurgle, as this complaint in the calf resembles diarrhoea in the human, and that of the worst type. I also know that this rarely ever failed to cure such cases.

This tree had another and occult property, with which the people in my country were acquainted; this was considered an antidote to demoniacal influences; or the ill wishes of bad people or black magicians.

A portion of this tree was considered fortunate to be kept in the house, also in the out buildings where the cattle were. It was a very common thing for farmers to make a wreath of the twigs of the Mountain Ash, to put about the necks of calves and other creatures, to preserve them from a complaint called the 'strike'. This was sudden death from a stoppage in the circulation. If a farmer wanted a rod to drive his creatures, he generally, if convenient, preferred a branch from this tree.

All this would now be called superstition, and simply the result of ignorance, which is at this day supposed to have abounded at that time; for how could it be otherwise, seeing there were no board schools then, nor parish and county councils to look after the people and spend their surplus cash! It is almost a miracle, one would suppose, that people could then live at all, and that in the absence of all our modern fads and schemes of plunder. But they did live, and the farmers prospered very differently to what they do at this day.

Whatever may be now called superstition, that superstition must have had a foundation in fact. Although these observances I have been alluding to might be characterised as being very ridiculous; in the meantime there is a truth which underlies this seeming folly which it will be my duty to bring to the height; and when I have made known all I am conscious of respecting this tree, it is possible you may feel more leniently disposed towards those men and women with their simple faith in the virtues of the Mountain Ash.

One night, while musing on my bed over a case of gout that had come under my notice, and for which I wanted to find an antidote on the soul plane; it was not long before the Mountain Ash came before my vision; and, in the meantime, before any further looking into the pathological property of this tree, my mind turned spontaneously on the cause and nature of this very terrible complaint.

I am fully aware that medical men are not quite agreed as to the primary cause of gout. It has been, I believe, taken for granted, by several of the old school, that it is the result of acid getting into the circulation; a kind of leakage of Uric acid which should otherwise have passed away by way of urine. Others of the profession do oppose this theory and affirm that acid has little or nothing to do, as a factor, in this complaint; but that it is a species of nervous affection, having characteristics peculiar to itself, which however depends on physiological idiosyncrasies promoted or sustained by habits in which the patient may have indulged; these habits being not necessarily confined to eating and drinking and that exclusivity. So far as I have been able to diagnose this complaint, by soul powers, I may say that there is much truth in both of the forementioned theories.

The chief mistake appears to be that neither have gone far enough, each of the parties have stopped at the portals. I will in this paper offer a few remarks which may be taken up by the profession, and applied scientifically, as they have the opportunities of doing so; by which the truth of what I herein write may be tested. In the first instance I discover an acid in the blood. I am not able to say what this acid may be, I do not think it is Uric acid, but may owe its existence to a combination of such acids as are in the food we eat, and the different liquids we drink. I know such combinations do afford the conditions for the development of an acid whose specific gravity is not so pronounced. I name this Neuro-toxic acid. When this is produced in the blood it dissolves, or destroys, the red corpuscles; because of this destruction, a peculiar lymph is generated in place of the red corpuscles.

This lymph is an acrid poison, yet not of so virulent a character as to take away the life of the sufferer at once, unless this lymph finds its way into some of those vitals parts of the body where its effects may be more speedily developed.

It is when connected with the nerve tissues that it does the greater mischief. It is very dangerous when connected with the organs of respiration. But when it reaches the brain, unconsciousness, or lasting idiocy, or sudden death, is the result. Many a supposed death from apoplexy may safely be attributed to this complaint. I can safely say that there is no medicine yet discovered that can effectively grapple with this malady. The Mountain Ash has come before me as having something, either negatively or positively, by which this poison may be nullified.

There is in the psychic nature of this plant an occult property which is negative to the pains of Gout, and which can attract to itself that virulent aura, and thereby eventually, take away this complaint. Whatever may be said respecting the supernatural, there is nothing after all outside nature. What is nature but the workshop of the Infinite! Everything in nature is the subject of a force, and is also capable of transmitting a force.

In the meantime, not that identical force which it receives; simply because each subject becomes a chemical laboratory peculiarly its own, where, by virtue of those mysterious, because complicated, appliances it does generate another force, unlike that which it receives. I will illustrate this matter: An Alkaline substance becomes the recipient of an acid. What is the force generated? A Salt or a Saline.

This Saline is unlike both of its factors. The Infinite has innumerable hosts of agents in this great workshop, and the subjects of his power are just as innumerable.

Each of these receive, through certain agencies, a force which comes from the Spirit Absolute, through the psychical. Seeing there is nothing higher than Spirit: God is a Spirit; there is nothing lower than what I call the objective material universe. And seeing these are all related, these are all within the domain of nature; where then lies the possibility for the supernatural? I have in the above remarks pointed out the basis of my Philosophy. I now proceed to particularise.

The Mountain Ash is a natural object, it exists in this Great Workshop of the Infinite. It is the recipient of a force. By virtue of this peculiar force, this tree is not some other tree. By virtue of this force the leaves of the Mountain Ash are serrated; the bloom is orange coloured; the berries, when ripe, are scarlet.

It is by virtue of this primal force that these same flowers have a perfume peculiarly their own; and that the juice of the fruit is intensely acid. In the meantime, none of these properties are to be found in the Primal force that is here related belongs, or pertains, to the outward; these are merely what can be sensed by us with one of three senses: The Sight, the Smell, or the Taste. But we have other senses which are prior and stand higher than the outward ones, and any power that effects the Soul senses is sure, ultimately, to affect the bodily organism.

For hundreds of years past, the sensitives of my own country have been able to appreciate an influence emanating from this tree unlike that of any other. In my vision I see an old Bard, sitting in the summer months, beneath the shade of the Mountain Ash. He falls asleep, and in that sleep he dreams that an Angel comes to him, and tells him to arise and convey to his neighbours, the glad tidings, that the tree under which he sleeps, and which is growing on their mountains, and in their forests, and, which all knew so well, had a virtue beyond that of any other plant.

That by making a Wreath of its branches, in a circle, and by hanging it up over the entrance door of the house, no evil influence, from witch or spirit, could enter that house.

That by placing the same about the neck of a creature, would prevent it from any accident, from evil wishes; that if they made a cross of two small branches, with the leaves, beneath the head in bed, that the sleeper would have true and important dreams, or revelations from God. But he gives no instructions about the berries, save this: that evil influences may come from many a source; that is: influences which would prove evil to persons under certain conditions, but which would be harmless in the absence of such conditions.

Some pride themselves in their supposed attainments in occult knowledge; just test that wisdom beside these researches which I am publishing under the Psychology of Botany. If your instincts, fail to conduct you into the spirit which pervades these Revelations. If you fail to appreciate these truths, cease hereafter to consider yourselves Occultists, much less Magicians.

Remember this; to become a Magician you must become a student of nature at first hand. I now come to deal with the inner nature of this tree: the influence which it receives from the Occult Spheres, and through those Divine Agents which stand graduated between the cause and the effect, is intensely blue; this falls on what I will designate an organic structure, that would of itself, subjected to the light of heaven, be a deep dull yellow.

By virtue of this soul influence of deep blue falling on the primal molecular substance, an influence is generated, which in its nature is that which pertains to a kind of hidden green! I will advance no further on these lines beyond this: that this tree, on its occult side, can attract to itself, and afterwards of retaining in itself, that poison, which I call, and that on my own authority, Neuro-toxic acid, which is the direct cause of the agonies of Gout.

<u>*Directions*</u> *After sunset, get some one to cut off a twig of the Mountain Ash, sufficiently pliable. Keep your design to yourself, if possible, and with*

this in view, do all with your own hands, if able. If the evil be in the lower limbs, put the twig around it for twelve hours, touching the skin.

Afterwards, cut the twig up in short lengths and bury them deep in the ground, and as the cut up fragments decay the complaint will pass away. But for those who cause their will power, and who are able to think of this tree in its absence, to such I give the following advice: After the hour of sunset, and before midnight, place yourself in an easy position as possible; fix your thoughts on this plant; possibly you may know where one grows.

Think of the Mountain Ash, and go over the following word eight times: AV-RL-TH-EI. Do this for eight days, and the pain will be no more.

Medicinal use of the rowan

Rowan is an anti-inflammatory tree used by the Physicians of Myddfai as a remedy for typhus. It is, however, combined with other herbs. It is also used as a laxative, and simply eating raw berries can cause severe sickness and diarrhoea. For this reason, it is usually served as a jelly or jam.

Rowan berry jelly

Prep: 15 mins
Cook: 60 mins
Resting time: 12 hrs
Total: 13 hrs 15 mins
Servings: 36 servings

Ingredients
4 pounds rowan berries
3 pounds apples, peeled, cored, and quartered
1 pound white sugar (for every 2 cups juice)
Water, preferably filtered

Steps in making rowan berry jelly

1. Gather the ingredients.
2. Put rowan berries and apples into a large pan or stockpot (there should be room for the berries to reach a good rolling boil and not be crammed in).
3. Cover the fruit with cold water just barely. Using medium heat, bring the fruit to a boil.

4. Reduce the heat to low and simmer for 20 minutes, or until the fruit is tender and soft.
5. Let the mixture cool for 5 minutes and place it, with a bowl underneath, in a jelly strainer bag overnight for at least 12 hours. It's crucial not to squeeze the jelly bag to extract more juice, as this will cause the finished jelly to become cloudy. Although it will still be delicious, it won't look as pretty.
6. Measure the juice you've collected and weigh the correct amount of sugar as directed above. Add the juice and sugar to a clean, non-reactive large pan or stockpot, and simmer over low heat for 10 minutes, until the sugar has dissolved.
7. Increase the heat and cook at a full rolling boil for 5 minutes, then test for a set following a setting-point test.
8. When the jelly has reached the setting point, pour into hot, sterilised jars, seal and label.

Enjoy with game meats, cheeses, or toast!

Modern medical research

The rowan tree (*Sorbus aucuparia*) has long been valued for both its nutritional and medicinal properties, and modern research is beginning to confirm its therapeutic potential. Rowan berries are rich in phenolic compounds, including chlorogenic acids, quercetin derivatives, anthocyanins, carotenoids, organic acids, and high levels of vitamin C, which contribute to their strong antioxidant capacity.[48]

Studies on sweet rowanberry extracts have shown significant radical-scavenging effects, with hydroxyl radical inhibition ranging from 16% to 25%, superoxide anion inhibition from 27% to 34%, nitric oxide inhibition from 25% to 31%, and lipid peroxidation inhibition from 8% to 13%, sometimes outperforming common antioxidant-rich fruits such as apples and mulberries.[49]

Rowan extracts also show promise as antimicrobial agents, with ethanolic extracts demonstrating inhibitory effects against *Staphylococcus aureus*, *Listeria monocytogenes*, *Escherichia coli*, and *Campylobacter*.

---

[48] Joanna Oszmiański and others, 'Bioactive Compounds in Sweet Rowanberry Fruits of Interspecific Rowan Crosses', *ResearchGate Preprint* (2023).
[49] Ibid.

Additionally, fruit pomace has been shown to suppress both Gram-positive and Gram-negative bacteria.[50]

Beyond these properties, rowan maintains a strong presence in traditional European herbal medicine, where the berries, often processed into teas, syrups, jellies, or tinctures, have been used to treat respiratory infections, colds, fevers, and bronchitis, as well as for anti-inflammatory, diuretic, and vasorelaxant purposes. Emerging research also suggests that polyphenols have antidiabetic potential, modulating blood glucose levels, improving insulin resistance, and reducing glycation.[51]

Magical uses of the rowan

As a protective tree, the history of rowan's use in magical practice is ancient. It can be seen in the use of rowan tree crosses documented in the records of the Pitt Rivers Museum in Oxford:

> *On February 28, 1893, three loops of rowan tree were donated to the Pitt Rivers Museum by Rev. Canon John Christopher Atkinson, from Danby Parsonage, Grosmont, York. (Accession Nos. 1893.18.1–3) These are now on display in Case 31. A—Magic, Witchcraft and Trial by Ordeal, located in the Court of the Museum. The records describe the rowan loops as amulets against witchcraft, but they also appear to have been prophylactic against ghosts, fairies, spirits, and the Evil Eye. All three loops are of different size, one measuring 70 mm at its maximum length (1893.18.1). Their provenance is stated alternatively as "England, North Yorkshire, Grosmont [Esk Valley]" and "England, North Yorkshire, Grosmont, Castleton".*

Amulets

Rowan can also be used in amulets. An amulet is believed to have some type of intrinsic power. Such things can be found in nature and may have been selected for their striking looks (e.g. the head of Medusa on a shield). The amulet can be used to protect people, animals,

---

[50] Kinga Rop and others, 'Antibacterial Activity of Rowan (Sorbus Aucuparia) Extracts against Foodborne Pathogens', *Antioxidants*, 10 (2023), 1779.
[51] Katarzyna Bucińska and others, 'Polyphenolic Composition and Antioxidant Potential of Sorbus Aucuparia (Rowan) Berries Compared with Other Fruits', *Open Agriculture Journal*, 17 (2023).

and property against various evils, including disease, magic, and death. Unlike charms, individual amulets function as prophylactics against multiple evils, depending on the owner's needs. The amulet literally 'turns away' (Gk *apotropaic* function), that is, of a particular evil and which gives the owner the necessary strength and protection through the amulet's function by 'shielding'.

In no way does the amulet cause the owner any harm. While some amulets require direct contact with the owner to be truly efficient, others function through their mere presence. For this reason, amulets are rarely, if ever, destroyed or hidden away, as is often the fate of magical charms.

Rowan tree crosses are said to have been made by the old men in Corgarff, Strathdon, Aberdeen. The cross's sacred shape and design is said to enhance the rowan tree's protective power. This type of cross amulet was therefore placed in every opening of a house, so as to keep witches out. On 1 August (*Lammas day*), such crosses allegedly had to be placed over all the doors at noon, in secret, by someone who did not stop and speak to anyone they met on their way.

According to a card catalogue for the 'loop' No. 1893.18.1, two of the three rowan tree loops had been placed as protection against witches on the railing of a certain Dr Alexander's house in Castleton, Yorks. On the other hand, the third of these loops had been fixed on a gate-spike before the church porch by a horseman who turned his horse thrice before setting each loop.

The power of the rowan tree appears here to have been enhanced through the performance of a magical rite. This type of practice is more commonplace when it comes to charms, since the efficacy of a charm depends on the actual rites and incantations that accompany its manufacture.

It might also be well worth mentioning that knots are generally ascribed to magical virtues in many parts of the world. Often they are considered spiritual fetters of sorts. Although the power of the knot may be maleficent, they can also act for the good of people and relieve them from evil. In Russia, many amulets derive their protective powers from knots. Crucially, though, the particular virtue of a knot only lasts for as long as the knot remains untied.

Personal experience with the rowan

My own rowan was rescued from the Forestry Commission tree stall at the Hay-on-Wye book festival. She was damaged, and one of her tiny twig branches was snapped. She was about to be discarded, but I took her in and repaired her with tape and lots of water. She is a potted tree with an open-bottomed pot, allowing her to communicate with her friends in the forest. She stands at the front of my house, guarding me and mine from the world.

Rowan comes through to me as a little girl, young, eternally small, fine-boned with long red hair. She is playful, joyous, possessive, and protective in the way children are.

She is a great teacher of magic, showing me how to use it to protect the environment and teaching me that the morality of magic is based on how much one is prepared to pay for their actions.

She teaches me that to influence others to fight alongside me, I must risk myself, my heart, and my emotional world. That scares me, and it's a passionate commitment I must make to her.

To me, rowan is Mars in Pisces; she has that joyous drive that pushes you to start new passions. She is infectious; once you contact her, she will be in your life forever.

Personal meditation with the rowan

She remembers how I protected her and now offers to return the deed.

She is big now, but still spindly, with leggy branches and whipping, wild-topped growth.

She comes as a red-headed girl, all legs and eyes ablaze. She stands legs apart, challenging with her gaze, eyes uptilted. Berries in her hair, all messy and flame touched.

Well, what will you do about it then?

No Karma or threefold law here, just the law of nature. Red in tooth and claw and, in this case, berry. Are you prepared to risk yourself for those you love?

Are you ready to give yourself for a cause?

Will you fight for me and the others?

For yourself?

She knows I am easily hurt and sometimes find people hard to deal with. And yet she pushes.

She knows I am at a stage where I just want to withdraw and rest. And she says, 'No, get on and do.'

She suggests with a smile and a raised eyebrow that I get off my old arse and stop moaning about the hard work. I chuckle, and as I walk back to my house, I can hear her laughter all the way.

168 BLACK PATHS AND GREEN CATHEDRALS

Fight for us and live

Rowan

The splash of colour in the green
Rising martial, gleaming in the sun.
Fighting to survive
Waving arms in the wild
Give me your heart that I might shatter it
In honour, give your heart to me
That I might open it and spill its contents on the land
Fight, fight with your being for the world
Fight for your place in the wild
I am your protection and your shield
Red hanging from my boughs
The wounds of previous battles Worn with pride. Fight on
For we must win

CHAPTER TEN

The Alder

The Tradition of the Alder

> *Gwern blaen llin,*
> *A want gysseuin,*
> *Helyc a cherdin, Buant hwyr yr vydin.*
>
> *Alder, front of the line, formed the vanguard,*
> *Willow and Rowan, were late to the fray.*

These lines are from the *Cae Goddau* or the Battle of the Trees, attributed to Taliesin, the famous Welsh poet. From the poem's order and the legends associated with Bran in the *Mabinogi*, alder has an obvious and essential role in Welsh culture.[52]

From the tree's connection to water and its ability to resist decay it becomes an active symbol of strength and endurance. It is therefore often used in the building of boats partly because of its ability to survive in water.

In the Norse tradition, the first man, *Ask*, was believed to have been created from an alder tree. The tree's association with water and its red sap also contributed to its symbolic nature and where in other cultures it is seen as a bringer of life, healing, and indeed of transformation.

Culpeper on alder

> <u>Description</u>: It grows to a remarkable height, and spreads wide, if the soil and situation suit. The bark is brown, and the wood redder than elm or yew; the branches are very brittle, and easily broken; the bark of the branches is spotted, yellowish within, and tastes bitter and unpleasant. The wood is white, and full of pith; the leaves are broad, round and nervous, and somewhat like the leaves of the hazel; they are indented, green, shining, and clammy. It bears short, brown angles, like those of a beach or a birch tree.
>
> <u>Place</u>: It typically grows near water or in moist, wet areas.
>
> <u>Time</u>: It flowers in April and May, and yields ripe seed in September.
>
> <u>Government and virtues</u>: It is a tree under the dominion of Venus, and of some watery sign or other, I suppose Pisces; and therefore the decoction,

---

[52] Marged Haycock, 'The Significance of the "Cad Goddau" Tree-List in the Book of Taliesin', in *Celtic Linguistics: Readings in the Brythonic Languages: Festschrift for T. Arwyn Watkins* (Amsterdam: John Benjamins, 1990), pp. 297–331.

or distilled water of the leaves, is excellent against burnings and inflammations, either with wounds or without, to bathe the place grieved with, and especially for that inflammation in the breast, which the vulgar call an ague.

If you cannot get the leaves (as in winter, it is impossible) make use of the bark in the same manner.

The leaves and bark of the Alder Tree are cooling, drying and binding. The fresh leaves laid upon swellings dissolve them, and stay the inflammations. The leaves placed under bare feet galled by travel are a great refreshment to them. The said leaves gathered while the morning dew is on them, and brought into a chamber troubled with fleas, will gather them thereunto, which being suddenly cast out, will rid the chamber of those troublesome bedfellows.

Medicinal uses of alder

The alder tree (*Alnus* spp.) has a long history of medicinal use, with modern research beginning to validate many traditional applications. The bark and leaves of *Alnus glutinosa* (black alder) have been used as astringents, anti-inflammatory agents, and wound healers, as well as for treating fever, rheumatism, oral and throat infections, and various skin conditions. These uses are linked to its rich phytochemistry, which includes tannins (15–20%), salicin (a precursor of salicylic acid), flavonoids, lignans, and phenolic glycosides. Recent studies have confirmed that alder extracts, particularly those from *A. glutinosa* and *A. nepalensis*, exhibit potent antioxidant and anti-inflammatory effects in vitro, as demonstrated by DPPH radical scavenging assays, which show significant activity. Traditional uses in Austrian herbal medicine, where *A. viridis* leaves are used in teas for fevers and infections, are also supported by laboratory-confirmed anti-inflammatory properties.[53]

Alder also shows promise as a natural antibacterial agent. Extracts from *A. glutinosa* seeds and *A. nepalensis* bark have demonstrated activity against bacteria, including *Escherichia coli* and Methicillin-resistant *Staphylococcus aureus* (MRSA). Additionally, *A. nitida* leaves

---

[53] News-Medical, 'Alder Bark May Be Great Source of Anti-Aging and Anti-Disease Natural Antioxidants', (2019) <https://www.news-medical.net/news/20191025/Alder-bark-may-be-great-source-of-anti-aging-and-anti-disease-natural-antioxidants.aspx> [accessed 25 October 2019].

have been shown to exhibit antidiabetic and hypoglycaemic properties in rat models, thereby reducing blood glucose levels and oxidative stress,[54] while *A. altissima* combined with $CaCO_3$ nanoparticles is being investigated as a novel nanotechnological delivery system for managing oxidative stress and Type II diabetes.[55]

Charubel on the alder

<u>The Alder tree</u> (Alnus glutinosa) *This Tree belongs to the birch family, catkin bearing and nut producing. The name for this tree in Cheshire is Owlar. I am not conversant with any other of those provincial cognitives which must exist. In the meantime, I do not think that any reader will fail to recognise this tree when I say that it grows beside brooks, stagnant pools, and swamps and boggy places, such are the situations where it thrives best; at the same time, it will not refuse to grow in dry places, and along road sides.*

This is one of those trees which is but little noticed, nor do I find but very few remarks respecting it in Botanical works, much less in Medical. When viewed as an object of interest on the outer plane it is not prepossessing; there is nothing attractive to the casual observer. It is dull, dark and ungainly, it does nor display the beauty and loveliness of other members of this family, such as the Oak, the Birch, the Poplar, and the Willow.

This sombre inhabitant of the morass appears fretful and peevish, and gives one the impression of one who is weighed down with grief and sorrow. One whose cruel treatment and general neglect had begotten in its outer nature the sourness of a misanthrope. Yet in spite of this repulsiveness, its inner bark is a good astringent for a relaxed stomach and bowels in man or beast, this I have proved for a number of years.

When a boy, I have known men whose feet had become chafed by perspiration and travel, gathering the leaves of the Alder Tree, and applying them to parts that were even bleeding at the time, who have afterwards resumed their journey without the previous agony.

You may thus see that this gloomy tree is not destitute of virtue, notwithstanding its morose appearance. But, as you already know, outward

---

[54] Katarzyna Bucińska and et al., 'Phytochemical Analysis and Antioxidant Activity of Selected Alder Species', *Plants*, 10 (2021), 2531.
[55] Muhammad Asad and others, 'Green Synthesized *Alnus altissima*-Conjugated $CaCO_3$ Nanoparticles for Diabetes Management', *Frontiers in Chemistry*, 12 (2024).

appearances are not always the safest of guides, this I shall be able to make clear to you in the present revelation of the Alder tree, in its occult and psychic properties as discovered by me on the soul plane. Here I find virtues of no ordinary type. In fact the word extraordinary would best comport with those ideas of which I am made acquainted.

An Occultist looking at the characteristic Sigil or Symbol of those occult forces possessed by this tree will not fail to discover what the Medical Botanist has not even the most remote idea of. I would here make a statement, and that at the outset, that this tree possesses magical virtues! and like the Mountain Ash can produce extraordinary results.

At the same time, the disparity between these two trees is very great. There is, in fact, a chasm between both that cannot be bridged over so as to unite or reconcile the both natures. The Mountain Ash is governed by Mars in the sign Leo.

The Alder tree is governed by Saturn in the sign Scorpio. You will readily see from these positions that the disparity is great indeed, at the same time, the magical power of each when employed on its own line, and within its own limits or sphere, is overwhelmingly great, grand, and glorious.

But, by way of caution, I would say that these strange forces must not be played with by the novice who may never have studied such occult laws, or who may not have graduated in the realm of soul. A person may make use of the bark, leaves, or roots with impunity, and that, in many instances with advantage, so long as he deals but with the outer tree and that on the material plane, a thing which anyone may use or cut up, according to his wants or his caprice, and that to his heart's content, as in such a case he is dealing with but the shadow.

However, this is not the case when he has to do with that world of realities, the domain of forces, the sphere of causation, that realm of celestial activities where causes are in continual operation, producing on the outer plane the phenomenal universe. It was the language of a primitive Christian, who was also an Apostle, that he "looked not at the things which are seen, but at the things that are not seen." At first sight this passage appears a self contradiction, for how could a person look at what is unseen? It simply implies the two-fold nature of man in his relationship with that twofold universe of which he is a part.

The inner world and the outer world; the inner sight and the outer sight; one adapted for the other. The Apostle addressed his pupils from the platform of an Occult Philosopher. The things that are seen are temporal or transient, but the unseen verities which are realised by the inner sight are eternal. Those forces which produce the outer phenomenal tree cannot be destroyed by the woodman's axe. It is true the tree as such, exists no longer, but the few remaining roots may shoot forth a new stem, or the seed from that tree may be carried away by some gust of wind to some genial spot where it may fructify by virtue of that power belonging to that plant, and thus once more, that power builds up or materializes another tree in the likeness of the parent tree which has been cut down.

It is to this same elemental plane I am directing your attention; and that sigil is the symbol of that power belonging to the Alder. For he who has wisdom may here find information above and beyond anything ever before published. All that is secret in this figure must remain so, as only the initiate will understand, and all I have met with are too clever to learn, so let each grope out his own way whilst I scatter abroad pearls of great value, and he who has eyes to see will gather them.

The Analysis I will, in the present instance, make known a few of those mysteries which are made known to me on the soul plane, and which I publish to the world. Glorious truths, such as this race has not heard of since the dawn of history. In the first place, the influence of this tree is of an isolating character. It tends to break up old associations, or old and intimate relationships, and that from the time a person or thing comes under this influence. Thus any excrescence, tumour, or any substance whatever joined psychically with this influence must inevitably pass away, from that moment you disconnect that substance on the soul plane it begins to operate on the physical plane, which will ultimately appear. In the second place, this influence has a reconstructing power; it not only effects the end of one thing, but also the beginning of another; on the one hand there is a death, on the other hand there is a birth. Thus out of apparent evil comes real good; or rather evil is succeeded by a good, yet not under the same conditions. Thus the influence of this tree if brought to bear on one's present life would disconnect one from the life of the past, and the mind would become disqualified for the occupations of the past, and, unless the mind had been previously prepared for soul work, the life of such a one would become a blank, he would appear to be nonprogressive and all would end.

This influence brought to bear on an imbecile would break up the old conditions, and the future would be altogether new. At the same time, there is no certainty as to how, or to what extent such may operate; hence it is not safe to apply this power to any other purposes than the destruction *of excrescences, tumours, long-standing ulcers, or any local complaint, but it would not be safe to apply such a force to the bodily constitution in any way.*

These powers may prove an advantage to the hermit or recluse, as its disconnecting influence would take away from him any remaining longings for companionship which might be lurking within his mind. And further, it gives or enhances that aversion to all that the world calls brilliant or glorious. It would render a person not only apathetic to the busy outside world life, but begets a positive hatred towards everything on this outer plane, so that there is nothing, however fascinating, that could prove a charm to one allied to this terrible power, neither music, or painting, or any of the productions of human genius; in fact, the very objects in nature, and of nature herself would prove unattractive or would be lost sight of. The great world itself like a moving panorama, recycles, it disappears. The grand unseen, the soul world alone opens to his eyes, his auditory powers on the inner plane become vibrant, as the winds from the unseen shore waft dulcet harmonies which awakes the aeolian harp within those mystic depths, and that for the first time. Such realities await the man who has lived, laboured and suffered whilst climbing the hilt, ever reaching out his hand to grasp the unseen.

Those influences of the Alder tree would apply beneficially to the character I have been describing. The word of invocation is CED-RAGEI- (pronounced Ked). This word should be repeated seven times, deliberately and with reverential feelings, having the Sigil before you at the time.

Magical uses of alder

The alder has a very long history of magical use. The alder-horse (in Finnish, *leppähevonen*) is known in folklore both in the eastern and northern parts of the country. This example is from Kiiminki and was written down in the 19th century. Still, it has been used by Finnish cunning folk over the centuries nonetheless.

When a new stable is built a horse is made of alder wood and for it a small stable. A blanket for the horse is made from a piece of the skirt of a woman who has recently given birth, and the eyes are painted on the horse with the woman's blood. Barley and quicksilver are placed in a basket and put in front of the horse. This alder horse and its stable are put under the floor of the new stable to give the horses good health and luck.

The *alder-horse* is an example of sympathetic, imitative magic. The actual horse deposit is replaced by the image of a horse. One might say that this is definitely a more economical way to make a horse deposit than if a real horse were used. The example is full of magical elements: firstly, alder is the preferred wood in Finnish magic, and the use of the skirt and blood of a woman who has recently given birth is also very typical for such a blood ritual.[56]

Alder appears to be a very popular wood in the Finnish magical tradition, being used for the handles of magical blades or for more complex magical workings, such as poppet-type workings known as the *Child of Alder*.

In Finnish folklore, a small (often 5–25cm of length), human-shaped wooden figure is used in magic, and has been called a 'child of alder' (Fin. *leppälapsi*), a 'child of the dead' (Fin. *vainajan lapsi*) or an 'alder man' (Fin. *leppäukko*).

An object of this kind would have been carved out of the wood in great detail and then magically adorned with hair and clothing.

On the other hand, in its simplest form, it could have been a figure made from a couple of attached branches, resembling a cross-like shape made to appear like a man.

---

[56] Cf. Sonja Hukantaival, 'Horse Skulls and "Alder Horse": The Horse as a Depositional Sacrifice in Buildings', *Archaeologia Baltica*, 11 (2009), 350–56.

Almost with no exception, the type of wood used in these figures was alder, and which by nature has a reddish hue, a colour very popular in both healing and baneful magic owing to its association with blood and to the tithe or clay used in the creation of Adam (Heb. *Adamah*—a play on the word for blood, *dam*).

A detailed description of such a wooden figure of the 'perfect human form', as it was said, was presented at the winter court sessions at Liminka in Northern Ostrobothnia in 1678. The object in question was an approximately 23 cm long, human-shaped figure carved out of alder. It had a felt jacket, socks, and slippers as clothing, the hair was made of hemp, and a scarf was tied with red thread.[57]

Personal experience of alder

There are many alder trees around my house. The land around me is wet, and alder trees love that. They are one of the only trees that have foliage, catkins and nuts all year. The dark green of the leaf extends all the way through from top to bottom and is shiny and silky to the touch.

To me, alder is a male spirit and is presented in a somewhat unusual manner. For he always looked off into the distance when he was talking to me, and was always standing in water. I never really got to see the face of this spirit; it was almost as if he was between me and whatever he was observing.

The alders are up the road from my house, and I have no alders in the backwoods.

Although he is masculine, he is also very emotional. When I meditated with him, my heart was overwhelmed by the emotions he exuded.

On one occasion, after the farmer had trimmed the hedgerow where he was found, you could see the blood sap oozing out of the severed twigs. Personally, I think this is why he is associated with the creation of mankind (e.g. *dam*/blood), and in the meditations that followed, he always had blood flowing and pooling in the water where he stood.

He is a liminal spirit, but not on the land nor in the water. 'Neither here nor there'. I think he corresponds to Venus, as mentioned in Culpeper, rather than Saturn, as noted by Charubel. The reason is that his transition from land to water plays a massive part in him. Protective,

---

[57] Juha Ruohonen, 'A Witch's Coin from Tervo', *Times, Things & Places*, 36 (2011), 344–57.

yes, but in an almost aggressive way. The land is what he protects, not the people on it, not because of any animosity, but because he feels the land and the people are the same. I think he would be *Venus in Scorpio* as his emotional and psychic side is awe-inspiring; he takes you deep into yourself. He opens your inner eyes and bridges those gaps in your understanding.

He is a complex person to get to know, partly because I suspect I have a complicated relationship with my own childhood and sometimes struggle to remember the positive aspects. Nonetheless, he would show me that my childhood had moments of profound joy and wonder, which led me to a more informed worldview of the present, where my relationship with the world became a more beautiful and awe-inspiring experience.

Meditation on the alder tree

It is wet where he lives; I sit on the wall over the outlet for the stream that runs out of the mountain.

He is all around me, sitting in the banks.

He is solid but quiet; he stands away from me when we talk.

He tells me stories of my childhood, things I have lost and given up to gain what I thought was valuable at the time. He makes me see the world with a child's eyes. It is not what it appears to be to adults; we see only the surface, the convenient, the ordinary. But to children, if they are allowed to truly see it before we push that sight out of them, it is alive. The world hums for them, glows at the edges, whispers secrets in the rustle of leaves and the curl of smoke. They see the magic we have trained ourselves to forget, until we teach them, as we were taught, to close their eyes to it.

He starts with love and the feeling you have as a child for the land, climbing trees, making camps and riding ponies.

The joy and freedom of being in the wild places. Of drinking water from old baths in the hedgerow put there for the cows, but you don't care if you are thirsty, and it tastes terrific.

He takes me along roads up onto fields I have not seen since childhood.

He digs things up from under the dirt and asks me to eat them. The taste of remembrance and smell of the earth. This place was a battlefield, so they said.

He speaks to me of my conversations with my pony, seeking advice and expressing my love for her. The joy of seeing rabbits racing in the fields. The wonder of being alone in the company of only the green and the spirits of the place.

He asks me to remember where I came from, before life, before my flesh enclosed my body and I feel the water of existence cover me, fill me, taking me back to the before. He says There is no beginning or end to life, just bridges between entrances and exits. We are connected by our presence here on this land. In this place. All things are connected by blood, air, bone, and water.

He always looks to the west and the sun setting, his back towards me.

His blood is in the water at my feet and flows away into the distance. Taking his spirit into the land.

Bridge the gap between and experience awe

Alder

Big man
Soft yet hard
Head turned towards the setting sun
Heart bleeding out into the water
Showing me a child's face
My own looking out onto the land
Drinking water from the cow trough
And eating nuts from the soil, where dead men lie after battle.
Those times when you saw the people around you clearly
The trees talked, and you listene
Your soul alight with the stories they tell
As we age, we need a bridge
Between one world and the other
Having made them two in our minds
By our own doing
Not understanding that we are one
One world with different people
One people with different lives
We are the same
You and I
Remember that child's face
That joyful knowingness of all

CHAPTER ELEVEN

The Birch

Birch tree tradition

Most traditional folklore around the birch tree has a direct association with renewal and growth. This factor is likely because it is the first tree to awaken after winter and the first to go to sleep after the autumn equinox.

In the Celtic tradition, the birch tree was associated with the goddess Brigid, the deity of healing, poetry, and craftsmanship. Birch also has an association with Arianrhod because of its silver bark. The birch tree represents renewal and purification. It has white bark, which symbolises purity, and its ability to withstand harsh conditions is seen as a sign of resilience.

The birch tree held a special place in Slavic traditions. It is associated with the goddess Mokosh, a deity of fertility, earth, and women. For this reason, birch trees are often used in traditional Slavic rituals and ceremonies and are frequently planted near homes to protect against evil spirits.

In the Norse tradition, the birch tree is connected to the goddess Frigg, who is associated with fertility, motherhood, and domesticity. The birch's use in brooms and besoms is linked to cleansing and, as such, magical rituals of purification.

In the Finnish tradition, the birch tree is highly regarded, considered a most sacred tree, and often associated with the goddess Rauni, who is said to inhabit birch trees and provide protection and healing. In Finnish, the birch tree is also used in rituals relating to the changing seasons.

Medicinal uses of birch

Humans have had an ancient relationship with the birch tree (*Betula* spp.), particularly through the use of its leaves and twigs, which were steeped in spring as a tonic when fresh vegetables were scarce. Birch leaf tea is a source of vitamin C, flavonoids, and other phytochemicals. Modern studies have validated its antibacterial efficacy in the urinary tract, as well as its anti-inflammatory and anti-adhesive actions against uropathogenic *E. coli* in bladder cells.[58]

---

[58] Elena Šavikin and et al., 'Evaluating Birch Leaf Infusion as a Functional Herbal Beverage: Anti-Inflammatory and Anti-Adhesive Activities in Urinary Health', *PubMed* (2024).

Birch leaves, containing flavonoids, tannins, and saponins, also act as diuretics, helping cleanse kidneys, stimulate bile flow, lower cholesterol, reduce blood pressure, ease fevers, and alleviate cold symptoms.[59]

Magical uses of birch

Birch is one of the 22 sacred trees in the Celtic Ogham, the sacred alphabet of the Celts. It is not surprising that birch functions ecologically in the UK the same it does in North America. Likewise, the theme of renewal, protection, and new beginnings is a consistent one. In the ogham, birch represents the letter 'B' and is 'Beith', being represented by a single line extending to the right of the line in the few. According to John Michael Greer in his *Encyclopedia of Natural Magic*, birch species are used interchangeably in terms of magical properties. Birch is represented by Venus in Sagittarius. Birch twigs were used for protection in traditional Western folk magic—a bundle of birch twigs placed along the edges of a property is said to keep away 'evil forces' and bad luck. Birch trees were tied with red and white cloth and put near stable doors to drive away elves (known to knot horses' manes and tire out the animals).

Birch appears in *The Long Lost Friend*, a 19th-century grimoire by John George Hohman, translated from the German and published in English in Pennsylvania. A charm, focusing on the restoration of the limbs or HOW TO CURE WEAKNESS OF THE LIMBS reads:

> Take the buds of the birch tree or the inner bark of the root of the tree at the time of the budding of the birch, make a tea of it, and drink it occasionally throughout the day. Yet, after having used it for two weeks, it must be discontinued for a while before it is resorted to again, and during the two weeks of its use, it is well at times to use water for a day instead of the tea.

Culpeper on birch

Description. This groweth a goodly tall straight tree, fraught with many boughs and slender branches bending downward: the old being coloured with discoloured chapped bark, and the younger being browner by much. The leaves at first breaking out are crumpled, and afterwards, like the

---

[59] Anđela Dragicević and others, 'Biological Activity of the Birch Leaf and Bark', *Natural Medicinal Materials*, 42 (2022), 89–105.

beech leaves, but smaller and greener and dented about the edges. It bears short cat-skins, somewhat like those of the hazelnut-tree, which abides on the branches a long time until growing ripe; they fall on the ground and their seed with them. Place. It usually grows in woods.

<u>Government and virtues</u>. It is a tree of Venus, the juice of the leaves, while they are young, or the distilled water of them, or the water that comes from the tree being bored with an auger and filtered afterwards; any of these being drank for some days together, is available to break the stone in the kidneys and bladder and is good also to wash sore mouths.

Charubel on birch

The Birch Tree (Betula) is as beautiful a tree as any which grows in a wild state in this country. Its silvery bark renders it a conspicuous object in arboreal scenery and may be distinguished at a considerable distance. It has been called by some "The Lady of the Forest." This is another species of tree that is native to this country. It is one that delights in most or mushy land and its domain extends the furthest north of any other tree or shrub. Being a plant so well-known by all, I need not waste time or take up space with further description.

Charubel's commentary on the birch

I now proceed to notice the planet and sign which are sympathetically related to the Birch: These are the planet Mercury in the sign Pisces. In the remarks I am making here and the revelation I am now publishing about the Planetary government of the Birch tree, as judged from a Stellar standpoint, I am not consciously expressing what any other person may have said or written. I am simply presenting what I find, regardless of the views of others; in the meantime, I respect the views or opinions of others, so long as they are the outcome of honest conviction, even if they may clash with my own findings.

Honesty of research and purity of purpose deserve our respect at all times, and that under every circumstance, so long as each one gives what he finds and that to the best of his ability. The Birch is a plant which partakes of the nature of Mercury and Jupiter combined. You may readily convince yourself of this by a studious and careful observation. Mercurial plants, overall, are tough, stringy, or fibrous, as is common in the hemp family; in this respect, the Birch bears some resemblance to the Hemp tribe.

There is yet another property in the Birch which bears a further resemblance to the Hemp, and that is, its stimulating power over the nervous system in general, and the brain in particular.

I know from experience that a decoction of the Birch branches, or what is better the inner bark, which is stronger than the branches, acts very powerfully as a stimulant on the brain. Indeed, as the result of my personal experience, I am forced to the conclusion that it very much resembles our grand old household beverage, the cheering cup of Tea. I have thought, and that repeatedly, that with a little cooking, and by the addition of some light vegetable aromatic, such as may be found in the cowslip bloom, or the petals of the white rose, that a substitute of a very invigorating character for the ordinary tea might be produced.

One thing I know, that the decoction of Birch alone possesses a wonderful power for clearing the head and sharpening the intellect at those times when a sluggishness seems to becloud the mental faculties; and as this is quite harmless, it would, if applied to, prove a boon to thousands.

In this regard, hemp, as a family, differs as it is not as safe; hemp is a laxative and diuretic, whereas the Birch is not a laxative but is an excellent diuretic.

There is another feature in the Birch which declares it a partaker of the influence of Jupiter; and this is its white and silvery bark. Jupiter is sympathetic with whiteness combined with brightness, hence block tin is a metal of Jupiter. Further: The sign Pisces is called the night house of Jupiter, and in judicial Astrology, represents moist or marshy land, and boggy places, but not what may be called filter places; and as the Birch seems partial to such localities, I think it must be an obvious truth that the sign.

Pisces must have some influence over this tree, for things or properties equal to the same thing must be equal to each other, in a certain sense at least.

These are a few of my outward evidences on the planetary government of this tree; I give them, as they may interest those who have some knowledge of Astrology. I sincerely hope that what I have adverted to respecting the natural properties of this tree may induce a few of my friends to a further investigation of such a matter. I would be pleased to receive from anyone who may have given some attention to this subject the results of their investigations.

I proceed to a further consideration of what I find on that plane with which I may be more conversant than those I am here addressing: I may give something of a very different character to what was afforded my readers respecting the Alder tree. I make this remark simply because Botanists place this tree in the same category as the Alder. Both yield catkins, and the leaves of both are very much alike, and are partial to like situations; further and beyond all this, there is some similarity in the property of the inner bark of each.

On the psychic plane there is a great dissimilarity, which I shall be able to point out alternately. The psychic Birch appears in the following garb: The colours are variegated; a blending of Yellow, Green, Pink, and Blue; neither colour very defined, but a shading of one in that of the other, as if the whole were blended into one; at the same time, not with the same effect as that produced by a similar admixture of colours by the art of the painter; in such a mixture there would be a blank indefiniteness, whereas in this psychic phenomenon each colour is distinguishable, yet one species of 'blend'. The apparent unity is real harmony, and the harmony is unity. The impression is that of softness or a mellowness, exceeding anything of beauty or loveliness on the outer plane of life.

Further, this tree on the psychic plane appears more symmetrically formed. It must be ever borne in mind that we must not expect such perfection of organisation on this outer plane as what exists on the inner.

The inner is the ideal; this ideal is the design of the Divine Architect! Here all is perfection! Yes this Divine ideal is the marvellous program given out by the grand First Cause—to those subordinates designated: 'Thrones, Principalities, and Powers'; and however perfect the design might be, the execution of that design, being entrusted to finite intelligences, who must develop with their work, cannot appear so perfect when under these outward adverse conditions as the ideal may be in its absolute condition. Thus it is, that there is a builder up of every tree, plant, and shrub. Yes, and this builder can paint as no mortal artist may be able to rival. There is a builder of the Birch tree.

The Symbol here given is the Occult Builder of the Birch tree. That Being who works by the ordering of Eternal laws. It is that being, or by virtue of his power that those special elements are collected together, and focalised in the form of this tree, and that according to the grand design and eternal ordination. Within his own sphere, this servant of the Most High is omnipotent. This servant manifests his wisdom and his power, and his intelligence, in that special outward manifestation; some living organic structure, where is found its ultimate on the outer plane. This Being is not dependent on the organic tree for its existence any more than electricity is dependent on the thundercloud for its existence.

> *The element called electricity is universally distributed, but its development to our sense of vision as a spark, or in the lightning flash, depends on certain conditions; and even these conditions are not the fortuitous mumblings of a purposeless fatuity. In nature, there is no such a thing as a movement of any kind, from the terrible collapse of a world, to the falling of a leaf in Autumn, but is the result of a Power; that power is combined with a degree of intelligence, an intelligence subordinate to a higher one. To me the whole Universe is like a hive, containing beings more active than the 'busy bee'. Yes, the Universe is one grand Pantheon, each chamber, each niche, each recess, is tenanted by god; each god is delegated with omnipotence in its own sphere; at the same time, the whole of these gods are controlled by one supreme head.*
>
> *Such was the most ancient creed of primeval man, at a period when he, in the outward form, could hold communion with these gods; in which sense it was he "walked with the gods," and as the result, he triumphed over the law of disintegration, of which he by nature was a subject.*
>
> *In the next place, I will make known to you those conditions to which the virtues of the Birch do apply. I feel a degree of certainty that there are among my readers, a few at least, who will feel grateful for this information.*
>
> *Restlessness, nervous irritability, accompanied with great anxiety. Direct your thoughts to the Birch, going over the word and Invocation here: AM-VEL-RAH.*

Personal experience working with the birch tree

The birches in my wood are old trees; they must be around 50 years old. But when they speak, they are young in tone and very androgynous. They don't have a gender in reality. They are quick communicators, mercurial and shifting.

They come to me in the form of young people and, most commonly, deer with spots and antlers. When they take the deer form, they are delicate in thought, and the communication is swift, like the wind against my face.

They are musical and intelligent beings, and I often wonder if Tolkien saw the birch as the elves in the woods. That is their feeling: age, but not aged; young, yet experienced.

They discuss the experience of mysticism, of beginnings, and the mind of creation. Their link with mushrooms is profound, and they can be a helpful ally when forming relationships with fungus for the first time.

They often talk about being the first and the oldest; they are proud of the ground they have laid, but there is also an understanding of the natural giving way to others. They are not long-term members of the community. They build for the future, paving the way for others who will move the community forward. Leaving links with the fungi to service the growing oaks and ashes.

They are the grandmother trees, looking to the future, imagining a time when they will no longer be, and arranging things so that a community can thrive without them.

Personal meditations with the birch

The deer approaches,

Quick, nimble, tiny and strong,

An intelligence directed solely at me, flickering through my mind's eye.

Iridescent colours shine, quickly shifting as the light shifts.

Knowing that I want to talk and learn, they come.

They lay their head down as I sit on the floor with my back to my old, old friend. They discuss with me the importance of beginnings and how they must be released to allow the community to move forward. The old must give way to the young. We discuss how humans have lost sight of this, with our obsessions with remaining young and the illusion of immortality.

The stifling of our young people by older, greedier minds. The imposition of the 'old ways' as the only ways. They talk about the impulse to know and to understand.

They talk about how our untrained imaginations cause us to fear shadows where there are none, slights and inadequacies where none exist.

They reveal fungal networks beneath the forest floor, shining with communication, breaking and reforming, as they search for new pathways and interactions. The need for community. The fungus clicks with joy at knowing that I see it. It shows me the recycling of all things into other things. One thing gives way to another; change is inevitable and propels life forward.

Give way and know immortality

Birch

Give way
Make space
For others to race
The course is not static
Life can become frantic
If kept in a box, where knowledge is lost and the way to the future
is blocked
Old ideas will fracture under the limits, community strains
With no growth and no gains
All vitality lost in the mirror of the vain
So create and move on
Live life and be gone
Give way to the young
Return to the earth
To be born anew.
Like a phoenix
From the ashes of yourself

CHAPTER TWELVE

The Holly

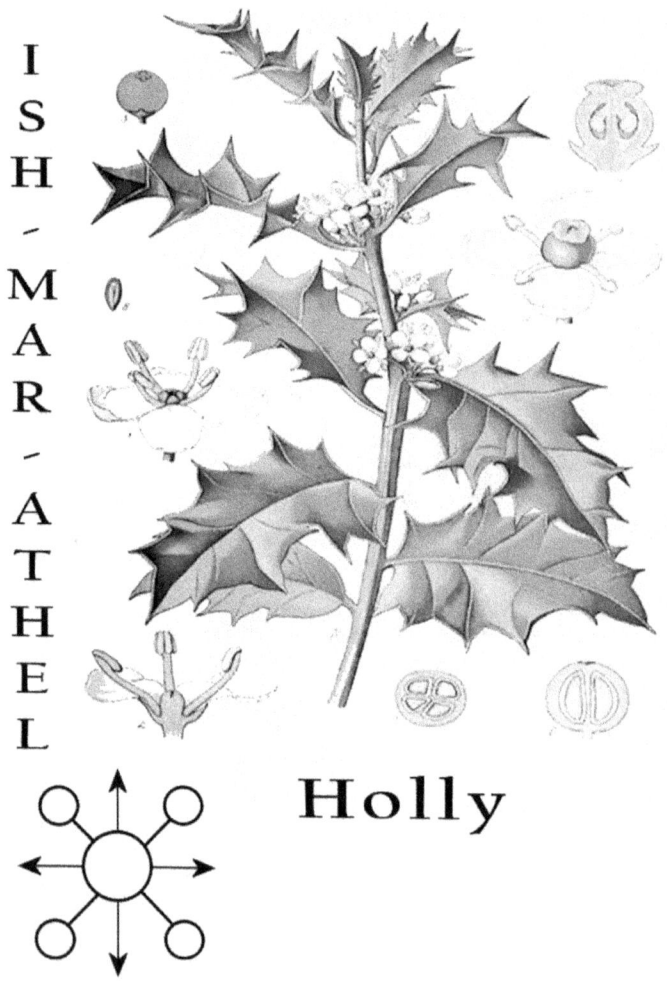

Holly tree in tradition

In the Celtic tradition, the holly tree is associated with the winter solstice and later with the festival of Yule. The holly is believed to have protective properties, guarding against evil spirits. It is also considered a symbol of fertility and rebirth, as its leaves remain green during winter.

In ancient Rome, the holly tree was associated with the god Saturn, who was honoured during the festival of *Saturnalia*, a time of feasting and gift-giving. Holly wreaths and decorations were used to celebrate this ancient festival.

In the Norse tradition, the holly tree is associated with Thor, the god of thunder. The spiky leaves of the holly were thought to represent Thor's protective qualities, guarding against lightning strikes and other malevolent forces.

During the Middle Ages, holly became associated with Christianity. Its sharp leaves symbolised the crown of thorns that Jesus wore during the crucifixion, and the red berries were linked to drops of Christ's blood. The holly's evergreen nature was later associated with eternal life and the promise of rebirth through Christ.

Branches of holly were hung over doors as a safeguard against demons, and the Roman author Pliny the Elder promoted the idea that hollies protected dwellings from being stuck by lightning.

However, while it is acceptable to prune a branch of holly here or there, the actual cutting down of an entire tree is definitely to be considered as bringing bad luck.

Culpeper on holly

For to describe a tree so well-known is needless.

<u>Government and virtues</u>. The tree is Saturnine. The berries expel wind, and therefore are held to be profitable in the cholic. The berries have a strong faculty with them; for if you eat a dozen of them in the morning fasting when they are ripe and not dried, they purge the body of gross and clammy phlegm: but if you dry the berries, and beat them into powder, they bind the body, and stop fluxes, bloody fluxes, and the terms in women. The bark of the tree, and also the leaves, are excellently good, being used in fomentations for broken bones, and such members as are out of joint. Pliny saith, tree branches defend houses from lightning and men from witchcraft.

Charubel on holly

<u>The Holly tree</u> (Ilex aquifolium) *I am sure that my readers would consider any attempt on my part to describe a tree so well known as unnecessary. The family of Holly is a small one and contains but a few species. The species of the Holly are natives principally of North and South America, the West Indies, and the Cape of Good Hope, and include a good many plants useful to man, the most celebrated being the shrub that yields the Paraguay tea extensively used as a beverage in Brazil and the adjoining governments.*

A few species from the Cape of Good Hope are esteemed for their foliage, and Prinos from North America are occasional inmates of gardens, but the only one generally cared for is the common European Holly.

*The tall and sturdy figure; the symmetrically conical form; the glossy and perennial foliage, with the fine effects produced by the innumerable leaves; and the gay bracelets of scarlet berries with which its branches are encircled all through the winter, may well have given preeminence to this beautiful tree as well as rendered it the most popular of British plants.—*British and Garden Botany, *by Leo Grindon.*

My sympathetic readings within the sphere of the Holly embolden me to pronounce that the Holly is an offspring of the Solar Rays or, as Astrology would denominate it, a plant of the Sun.

It is for this occult reason that from time immemorial its branches have been used to decorate the homes of the people at Christmas-tide. It is about this time that the Sun enters Capricorn, which is that ascending point when the Sun begins to climb out of the darkness of winter, when the days begin to lengthen, and when it may be truly said the Sun is borne, hence it has always been a season attended with hilarity and mirth. This was the custom of mankind long centuries before the advent of Jesus into this world. Some of the early Fathers of the church, when the true Christianity had become corrupted, in order to adapt their religion to the so-called heathen, adopted their festivals, more especially Christmas and Easter, and gave it out that the former memorialized the birth of the Son of God and the latter was in commemoration of His crucifixion which is the time of the Sun's ingress into the Equinoctial sign Aries, when the lord of day is on the cross such is at that time his position in the heavens. But the true religion of the gospel of Jesus Christ has nothing to do with any of these festivals, they are innovations brought about by the apostasy. There is no command in the New Testament to keep the Sabbath day, and much less is you chose to but not as an institution. There is an injunction to hold a Christmas day. If keep up that season for festivities and mirth, do so, Christian rite, for it is a slander on the Christian.

The Holly on the Soul Plane

I have not yet discovered a plant or tree whose identity is so conspicuously manifest on the soul plane as the Holly. It bears a striking resemblance to its earthly type, the only difference being, that where the berries are in the earthly, from thence proceed tiny jets of golden light of starlike shape, the centres being of fiery red, which conveys the idea of a glow of heat; like so many outlets from one fountain of Celestial Fire which occultly permeates the more inner portions of this wondrous tree.

The Holly applies sympathetically to a greater number of ailments than any other tree I have yet described. It is also generally fortunate, and particularly so if planted on the south side of the house, and more especially is this so if the inhabitant be in sympathy with this tree. This may readily be discovered by attending to the following remarks: The holly is solar in a primary degree, consequently, solar people would derive a benefit through this tree being on the south side of the house.

This tree applies favourably to all persons and things under Mars, because when the Sun enters Capricorn, he enters the exaltation of Mars. In such a case the Holly should be on the west side, as this is the position for Mars people and Mars work. It applies favourably to all under Saturn, as Capricorn is the house of Saturn. All persons born under Saturn may derive benefit from the Holly, and should have it placed on the east side of the house. Thus there are three classes of people who are specially benefited by the Holly:—Sun, Mars, and Saturn. The other planetary people may derive a benefit indirectly, that is, through the agency of others, but not directly. This is a point which any Astrologer may discover.

The Psychic Cure

The complaints for which the Holly is the antidote are as follows: Great heat in the throat; constrictive or cramp-like feeling about the region of the heart; spinal affections; virulent pains in the head, accompanied with great heat; parched tongue; intermittent fevers; and delirium tremens.

The secret of all such cases as I have pointed out being curable by the Holly, is this: that element which is the cause of such complaints, is out of its place in the human body, but it is in its place in the Holly. By you thinking about the Holly you become in sympathy with it, and the Holly being negative to that element which is the cause of your suffering, attracts this element to itself, as this is needed by the Holly. It is *thus that every plant and tree that grows on the face of this earth absorbs some one element, which by its multiplication out of its own species, generates some characteristic complaint in animal life, which its species, if sufficiently numerous, would attract spontaneously, but which is often absent from the place where it is required. This defect may be met by the mind directing its energies to that plant or tree independent of position.*

Teach this to your children, and who discovered this divine method of cure. It is the duty of all to publish these principles broadcast; seek to instruct your friends, there is nothing to be ashamed of and you will significantly benefit yourself by so doing. "He who waters shall he watered."

Personal experience working with holly

The holly trees in my woods have grey/white bark. They are short trees that live in the undergrowth of the larger oaks and ashes. They twist and turn through the woods, making progress, and it is a real challenge on occasion. They grab your hair and clothes, stopping you from proceeding.

Holly came to me as feminine, which I did not expect; with all the traditional mythology that is associated with the 'holly kings' and 'green men', I fully expected an experience of male potency.

What I received was in fact the complete opposite. She was small and always carried a lantern. She was literally a light in the darkness. She talked to me about the role of sharing. She talked about sharing light, warmth and the value of community. Nothing alive can survive without the life energy of others, both for companionship and nutrition. Feeding both the soul and body. She made me face the selfishness of humans, that the primary thought for most was only for their own existence, with no understanding that without the wonderful web of livingness that supports us, we would simply cease to be.

She spoke about the shared experiences of life, love, joy, and the sharing of them, as well as the illusions that our society has promulgated and the negative effects that these have on us all. She also talked about the illusion of the importance of separateness.

Meditation with holly

In the darkness, she walks in a pool of light from her lantern.

The air smells of spices and sweet fragrances.

She is small and has white hair but is young in the face and has a welcome that you cannot refuse.

She is the warmth at Yule, that connection with others that only appears with shared experiences.

She speaks quietly, and you lean in to hear her whispered words. She talks of community, her voice rising in passion when she feels she is not being understood.

The survival of all is the survival of one; no life is taken lightly, but instead given with the understanding that for all to survive, sacrifices must be made.

By giving your heart, your body, and your soul, the spirit grows.

Living in a community means limiting your own growth in service to the growth of others. To lie down and give yourself to the good of the group to put others before yourself instead of pushing on for our own ego service.

To understand the web between all things, the balance it achieves, the joy of the resonance of the tribe.

And she says, 'Hear me well, standing alone in the dark, shining bright only for yourself, that is not my way.'

That is the path of the martyr, and though it may look beautiful, it denies the greater gift.

No. Pass the light. Share it. Hand to hand, heart to heart, so none stand in shadow. When all are alight, we are stronger, brighter, laughing as one. This is my teaching. This is how you grow with me, rooted, woven, alive in the great livingness of the earth.

200   BLACK PATHS AND GREEN CATHEDRALS

Come together, and the Light grows

Holly

Dark the land around us
Light shining in the black
The sound of singing voices
Pulls joyful spirits back
Community is life here
And sharing is the way
Of body, soul and spirit
That all may find their way
One does not enlighten
One does not survive
Only working all together
Allows the tribe to thrive
And not just tribes of humans
But planetary wide
We are all in this together
Holding light
To share
And live

CHAPTER THIRTEEN

The Hawthorn

Hawthorn

The hawthorn in traditional folklore

The hawthorn tree (*Crataegus monogyna*) holds a special place in the mythology and folklore of various cultures throughout history. Commonly known as the *May tree*, the hawthorn is often associated with magical properties, protection, and deep symbolism associated with the 'white'. The hawthorn is linked to the May Day celebrations in many cultures, and is particularly associated within the Celtic and European traditions. On 1 May or 'Beltane', we gather hawthorn blossoms to decorate maypoles and dance around them to welcome the arrival of spring. It is the tree's white blossoms that are seen as a symbol of fertility and that of new beginnings.

In the Celtic tradition, the hawthorn is sometimes called the *fairy tree* or *faerie thorn*. It is believed by some that the trees are inhabited by fairies and other supernatural beings. Disturbing or cutting down a hawthorn tree is thought to bring extreme bad luck and can cause the wrath of the fairies. I have a clear memory of my gran giving me a massive telling-off after I picked a bunch of may blossom. I remember my dad calling it 'Bread and Cheese'. It is customary in some places to leave offerings at these trees to appease the fairies and invoke their magical protection.

The Hawthorn is often seen as a protective tree in various cultures. It was commonly planted near homes and fields to ward off evil spirits and any negative energies. The thorns on the tree were thought to be a deterrent to malevolent beings, both physical and supernatural. In some cases, people would even make amulets or charms from hawthorn wood or its branches to carry with them for protection.

A well-known folklore is the saying *'Cast ne'er a clout ere May is out'*. This seems to have been misinterpreted to mean the month of May, but it almost certainly refers to the opening of the hawthorn flowers, rather than at the end of the month.

In DragonOak, we do not celebrate Beltane until the *white blossoms of the hawthorn are out*. We live in Wales, and it is cold here. We sometimes have long winters, so this means that we celebrate anywhere between 1 May and the middle of the month, depending on the local weather. We believe it's essential to work in harmony with the land and the trees, rather than imposing a calendar on them.

Culpeper on hawthorn

It is not my intention to trouble you with a description of this tree, which is so well known that it needs none. It is ordinarily just a hedge bush, although when pruned and dressed, it grows to a tree of reasonable height.

As for the Hawthorn Tree at Glastonbury, which is said to flower yearly on Christmas-day, it rather shews the superstition of those that observe it for the time of its flowering, than any great wonder, since the like may be found in divers other places of this land; as in Whey-street in Romney Marsh, and near unto Nantwich in Cheshire, by a place called White Green, where it flowers about Christmas and May. If the weather be frosty, it flowers not until January, or that the hard weather be over.

<u>Government and virtues</u>. It is a tree of Mars. The seeds in the berries beaten to powder being drank in wine, are held singularly good against the stone, and are good for the dropsy. The distilled water of the flowers stay the lask. The seed cleared from the down, bruised and boiled in wine, and drank, is good for inward tormenting pains. If cloths or sponges be wet in the distilled water, and applied to any place wherein thorns and splinters, or the like, do abide in the flesh, it will notably draw them forth.

And thus you see the thorn gives a medicine for its own pricking, and so doth almost everything else.

Medicinal use of hawthorn

Hawthorn is famous for treating cardiac and circulation issues. There are several ways of making herbal remedies here.

<u>Teas</u>: To create teas (infusions and decoctions) from the hawthorn, use the leaves and flowers or de-seeded berries. For a strong infusion, pour boiling water over the leaves and flowers, steep for 10–20 minutes, and drink. For the berries, bring water to a boil, add the berries, and boil covered for at least 20 minutes (depending on whether they are whole or smashed before drying).

<u>Syrups</u>: Chop hawthorn and cover with 1 quart of water. Boil this for about an hour, then strain the berries. Boil it down to 1 cup, then add sugar. My gran would make this in combination with rose hips. The old Welsh name for it was hips and haws.

<u>A tincture in vodka</u>
Place fresh leaf, flower, or berry (fresh or dried) in a glass jar and cover with vodka.

Let sit for at least 2 weeks. Strain through a small-bore sieve. Store the tincture in a glass jar in a cool, dark place. Take 30–40 drops three to four times a day.

For menopause, the intake of a herbal decoction might be beneficial for managing hot flushes experienced by women during menopause, as it can occur as a part of ovarian ageing.

Pour a cup of boiling water into a teapot.

Add 1 teaspoon of dried hawthorn berries to the water, cover and let it steep for 8 to 10 minutes. Strain the mixture, sweeten it with sugar or honey and drink.

Similarly, you can prepare a healthy drink with the dried leaves and flowers of hawthorn by boiling them in water for 15 minutes, followed by steeping for 15 to 30 minutes.

Charubel on hawthorn

Sadly, there is no commentary from Charubel for this wonderful tree, which is surprising considering it is a widespread tree throughout Wales.

Personal experience of the hawthorn

There are lots of hawthorns in my area; farmers use it as a hedge to stop the Welsh sheep wandering, as they are apt to do.

Hawthorn was a strange tree for me; she was twofold. The first tree I worked with was the one that lives out in the woods. I first started working in the autumn, and she had red berries, not so many, as the birds in my wood had been very active and sat fat on the branches of the birch she lives next to, shouting at me to leave the berries alone. She is an old tree with very few leaves and lovely, rough bark; when I sat with her, she was very welcoming, and there was an excited quickening of my heartbeat when I made contact with her. She came to me as a wrinkled, white-haired old woman with a youthful eye that held you in her thrall. She spoke to me about the end of life, the rejoining of the earth, and the gifts you give to future generations.

The second tree I met was in the hedgerow by the house. Her spirit was much younger and more like the May queen I had expected.

She was lithe and young, and came with red skin and hair. She always appeared with white blossoms in her hair and very red lips.

She was flirtatious and sexual, and I felt she was ready to explode with creative energy.

The old woman and the young woman were essentially the same spirit. I could feel the young in the memory of the old and the old in the potential of the young.

She spoke to me about life changes and how we navigate the various cycles of our lives.

This sounds like it was oriented towards death, but it was not that feeling; it was about the cycles of sex and reproduction.

Meditating with her, I reflected on how my body had changed over the last eight years. I am now 58 and have gone through menopause. I am a Scorpio sun and an Aries moon. One of the things I found most challenging to deal with during the changes from the menopause was the changes to my body, which made me a less sexual being.

I had always used my sexuality as a weapon to both get what I wanted and persuade people to think the way I did. I could no longer do that. My body had now become that of an older woman. The things I mainly noticed were the strangest; obviously, my breasts were lower than they used to be, but no one told me about how they would change to feel and touch, and my skin was less elastic, but no one spoke about the changes to your jaw and that your face shape changes. Strangely, I missed my menstrual cycle. It was a very odd feeling knowing that I have one beautiful daughter I love beyond all recognition, but having the choice to have another child taken away, even though I didn't want another child, to lose the avenue of possibility was quite hard to deal with.

Working with the hawthorn made me think more seriously about the fact that we, as women, are almost frightened to discuss things like menstruation, childbirth, and the menopause, for fear of revealing unpleasant facts. Yet, we all have to go through it at some point. It is a little bit like the taboo around discussing 'death'. Aspects such as these are indeed common in all life, and yet many people don't discuss or engage with these important subjects.

One of the most beautiful things about working with hawthorn is that she showed me the importance of connecting to our past selves. For me, as a young girl, this resonates deeply. It is simply that that energetic little child, running wild through the countryside, is still there.

And although my body is now very different, I understand that it is an honour to make contact, recognise, and literally ride with my inner-self, for this is an integral part of me and one through which my relationship with my being provides a link whereby I can experience the wonders of the world around me.

Personal meditation with hawthorn

She came to me as both old and young. Two distinct versions of the same spirit. The older showed me the children sprouting around her, yearlings of them waist-high but growing by the day. Her gift to the future. The younger talked about the urge to reproduce, to take joy in the sexual act, to know that her beauty shone through. The older one smiled and said that beauty, while wondrous and powerful, was transient and that humans are unforgiving of the life cycle. Hard on themselves, almost as if growing old was a failure of will. The young also shared that we don't sit with our youth and enjoy it enough, constantly living in the future or the past, but never in the present. Never feeling the joy of running until we look back on it in a nostalgic review. Always looking forward to who you will be next year or when you are 30 and suddenly your life has gone without you participating.

You are as young now as you will ever be, she says. Why waste it concerning yourself with what was and what is to be. The old talked of the privilege of life extended by looking at the world that you helped create with pride, and the younger talked of walking forward with that knowledge and improving on its foundations. When they leave, they always go arm in arm, at peace with each other, the future with the past.

The end is in the beginning and the beginning in the end

Hawthorn

Two-faced tree
Of beginnings and endings
Showing the way to live in the now
A little girl running in old shoes still runs as fast in her mind
The rising of red in her passion gone
The white comfort of the wisdom growing
Her berries open and ripe for the picking
Where white flowers bloomed
Berries hang heart strong
Transient beauty
Returning again and again
In different blooms
For different uses
Remembering to be young at any chance
Loving life's slow fatal dance

CHAPTER FOURTEEN

The Blackthorn

The blackthorn

The Blackthorn is often associated with darker and more mysterious aspects of folklore. Its thorns and dark berries contribute to its reputation as a tree with protective and magical qualities. In some traditions, it is considered a tree of enchantment, often found in the heart of forests or near mystical places. For this reason it is deeply connected to witchcraft and protection. Its thorns were believed to possess warding-off properties to keep away evil spirits and negative energies. Wands that are made from blackthorn wood are thought to have potent magical abilities and are used by practitioners of witchcraft.

The blackthorn's appearance in winter, with its dark, spiky branches and early white blossoms, has led to associations with death and rebirth. It is often seen as a symbol of transformation and the cyclical nature of life itself. In some traditions, the blooming of the blackthorn is considered a sign that winter is ending. Blackthorn is associated with the festival of *Samhain*, with its remembrance of the dead and the veneration of the ancestors.

In folk medicine sloe berries are used to make remedies such as jams, jellies, and tonics. The berries are rich in antioxidants and vitamin C, which might have contributed to their use for boosting the immune system and more generally for one's health.

Similar to the hawthorn, the blackthorn was often believed to be inhabited by fairies and other supernatural beings. Cutting down or disturbing a blackthorn tree was considered bad luck, as it could anger the spirits within it. The blackthorn was also used to mark boundaries between lands and properties, possibly owing to its thorny and protective nature. Indeed, in Christian folklore, the blackthorn tree has acquired a somewhat sinister reputation, often being associated with witchcraft and malevolent practice.

The thorns are also used in rituals, such as cursing and the 'binding and blasting of spells', where the intent is to cause harm or misfortune to others. The 'black rod' or 'black rod wand' is made from blackthorn and is often connected with malevolent magic. These rods were believed to have been used by witches to direct their harmful intentions and to place curses on others. The thorns on the end of the wand are seen as a tool for focusing and amplifying dark powers.

Excerpt from the Witches Museum concerning Iain Steele's blasting rod

<u>Size</u>: 890 x 80 x 70

<u>Physical description</u>: Blackthorn blasting rod—three blackthorn stems (cut from where a single branch is divided into three stems) twisted together into a helix and joined at the end with a short length of black tubing. Painted black and varnished. Tied with dark red ribbon.

Information from the collection of magical objects that belonged to Iain Steele:

> Blackthorn grows around holy wells and so is associated with access to the Otherworld and its protective powers. It is also associated with magical protection because of its sharp thorns (just as pieces of coral resembling animals' horns are used as amulets). Its use for blasting comes from the fact that attack is sometimes the best form of defence. (Many thanks to Mike Runa for this information.) According to Michael Howard, this rod 'is old and may have come from Norman Gills or some other Old Craft source'.[60]

Poppets are small dolls or effigies that are often used in sympathetic magic: and where what is done to the doll is magically reflected on the target. The blackthorn's thorns were said to be used by witches to pierce the effigy or poppet, thus causing great harm or misfortune to the individual the poppet represents.

---

[60] https://museumofwitchcraftandmagic.co.uk/object/wand-rod/.

Folklore surrounding the blackthorn also includes stories about the involvement and conjuration of the Devil. In medieval times, it was believed that the Devil would prick his followers' fingers with the thorns of the blackthorn tree as a mark of their allegiance. Sadly, both witches and heretics were at times sometimes burned on blackthorn pyres.

The Irish cudgel, the *Shillelagh*, was made of a blackthorn root. Historically, the shillelagh was used as a walking stick and a weapon for self-defence. Travellers could easily carry it and it served as a form of protection in more rural areas. The shillelagh's reputation as a formidable weapon was due to its dense and knotty wood, as well as the weight and balance of the club. One tale that made me smile was about a witch who carried a shillelagh simply to 'make mischief in the community'!

Culpeper's commentary

Slow bush description: *This is a bush, or small tree, whose tough branches are full of hard sharp thorns, sending forth its white five-leaved flowers early in the spring, before the leaves appear, which are small and oblong, finely indented about the edges. The flowers are succeeded by small round fruit growing on short stalks, green at first, but when ripe of a fine purplish black colour, of a rough sour austere taste, and not fit to be eaten till mellowed by the frosts.*

Place: *The sloe-bush grows every where in the hedges.*

Government and virtues: *This is a Saturnine plant. The fruit is chiefly used, being restringent and binding, and good for all kind of fluxes and haemorrhages. It is likewise of service for sore mouths and gums, and to fasten loose teeth. The juice of sloes being boiled to a consistence, is the Acacia germanica, of the shops, which is now-a-days made use of instead of the true and put into all the great compositions. It is of a darkish colour on the outside, and reddish within.*

It is the juice of this berry that makes the famous marking ink to write upon linen; it being so strong an acid that no other acid known will discharge it. A handful of the flowers infused is a safe and easy purge, and taken in wine and water is excellent for dispelling the windy colic. The bark reduced to power, and taken in doses of two drachms, has cured some agues. The juice expressed from the unripe fruit is a very good remedy for fluxes of the bowels; it may be reduced by a gentle boiling to a solid consistency, in which state it will keep the whole year round.

Charubel's commentary

Charubel provides no commentary on the blackthorn, and this makes me pause to consider whether Charubel, a Christian minister, would refrain from commenting or working with this tree given the mythology surrounding the crown of thorns and its connection to Christ.

Personal experience with blackthorn

While hawthorn had taught me how to live in the now, blackthorn taught me how to use the power of my voice, how to value boundaries, and more particularly, how to say NO!

She was a very awkward tree to work with; where I live, she is not really a tree, but a bush, and is primarily found in the hedgerow, where she is part of the farmer's defences against wayward animals and invading foxes.

When I first approached her, I wanted to create a wand to make a blasting rod that would help me work, aiding the ash tree in its fight against the fungus that was killing it. When I first approached her, I was given the most emphatic "no" that I have ever felt. There was no doubt about it. I couldn't ignore the answer, and I honestly think that if I had ignored it, I would have paid a very high price, both magically and physically.

I therefore moved on to find another blackthorn about half a mile away. Now in Wales, we use the technique of hedge-laying or coppicing, which partially cuts the trunk of small, pliable trees and lays them down horizontally to allow them to make a thicker, more efficient hedge (good for really determined sheep). As I approached what I thought was the second tree, I heard a distinct voice say, "it is still me and the answer is still no". I gave in and eventually returned to do some sitting work with the blackthorn to improve our relationship.

Blackthorn is Saturn in Capricorn, and for this reason, she is cold, unemotional, and astute.

She comes to me as an old woman with a wrinkled face, long fingers, with long nails. She rarely smiles and will often smack you, taking the price for the knowledge she has and is willing to impart.

Personal meditation with blackthorn

She comes reluctantly, only after I have spent a considerable amount of time asking her to come. She discusses how humans often believe they can do anything without consequence or consideration for others. That unfettered thought is dangerous, and the push for eternal growth will end badly for us.

She has no worries about her survival; she has been through much worse situations and has emerged stronger. She does not care that we won't survive; instead, she looks forward to that day, knowing that the green would thrive without us.

She is an angry spirit with tremendous power. Her anger is directed at our lack of understanding, our inability to see past our own needs, and our fear of upsetting the status quo. Yet, it is a statement of the obvious. She feels that we are stuck in the mire of illusions created by the inappropriate use of our power. And we are cowards for not attacking what we know is not real for fear of looking foolish in front of our peers.

She is loud and verbose and pokes me when she talks. She pushes me to speak and not to silence myself through self-imposed doubts. I talk to her about imposter syndrome, and she laughs in my face; what a joke and an ego! As if the universe thinks about you enough to worry about you looking stupid

Grow up, little girl, and get on with it. If you want to be a leader, then fucking lead!

The ultimate power is in your own voice

Blackthorn

No Girl No
When I say go, please go
You pretend that you listen; you pretend that you know
But the emperor stands waving his balls in the wind
and complicit in madness; you nod yes—it's a win!
He stands naked in gold dress of his imagined success
And you applaud like the sheep that you are.
What are you thinking? What are you doing?
Causing pain and destruction wherever you go
Clean up your mess
And while I confess
I will laugh when you're gone
To the place you belong
In the hell of your human creation
For we will go on, the green and the strong
Unless you remember
Unless you recall
That illusion reality makes
Point and Laugh at his low-hanging fruit.
Raise your head from the crowd
And say no, say it loud.
Stand apart from the fools who follow
Because you know that the truth they know
isn't True.

CHAPTER FIFTEEN

The Willow

Willow

In many cultures, the willow is associated with femininity, intuition, and the moon. Its long, flowing branches are often seen as representing the feminine principle. In the Celtic tradition, the willow is associated with Cerridwen and the lunar cycles, which symbolise transformation from one to another.

Willows are often found growing near bodies of water, and their association with rivers and streams has led to connections with emotions, intuition, and the subconscious. The willow's presence near water has inspired several magical cultures to view it as an interstitial bridge between the physical and spiritual worlds.

With its drooping branches, the weeping willow has long been associated with sorrow, mourning, and grief. In art and literature, the image of a weeping willow tree is often used to convey a sense of sadness or loss. In some cultures, its mournful appearance has made it a symbol of remembrance.

Yet despite its associations with sadness, the willow tree has also been linked to healing and protection. Willow bark contains salicin, the natural compound from which aspirin is derived. Traditional healers therefore use willow bark to treat pain and inflammation.

Medicinal uses of willow

White willow bark has been traditionally used as a natural remedy for pain relief. As noted, its salicin content can help reduce pain such as that caused by headaches, muscle aches, menstrual cramps, and joint pain associated with osteoarthritis and other inflammatory conditions. The anti-inflammatory actions of salicin in white willow bark make it beneficial for conditions such as arthritis, back pain, and several other inflammatory disorders.

White willow's bark can lower fever and has saved thousands of people from death by inhibiting the production of prostaglandins, which are chemicals that promote fever.

Modern research

The willow, or *Salix*, is more than a tree of moonlight and myth; modern research is beginning to catch up with what our ancestors sensed in their bones. Its roots are not only anchors in wet soil but also excellent absorbers of toxins, drawing mercury, cadmium, and lead from the earth, holding them quietly within wood and leaf. *Salix caroliniana* has been shown to sequester mercury, while fast-growing willows like *S. viminalis* and *S. alba* pull heavy metals and even oil-based toxins from polluted ground, healing the land as they grow.

Yet the willow is as much a healer of flesh as of soil. Its bark, rich in salicin, flavonoids, and phenolic acids, does more than echo aspirin; it eases pain, soothes inflamed joints, and quiets the ache of old injuries, as confirmed by modern clinical reviews, even if science still calls the evidence 'low certainty'. Beyond the bottle and the pill, willow retains its old green magic; farmers now use its aqueous extracts to coax life into other plants, encouraging blackberry cuttings to root and thrive. She is still, as she ever was, a tree of healing, turning poison to medicine, sorrow to relief, and death-heavy soil to green promise once more.

Magical use

Willow trees are often linked to the moon, the feminine, and emotional energy. They are seen as symbols of intuition, sensitivity, and emotional fluidity. Willow is often used in rituals and spells that involve working with lunar energies, enhancing psychic abilities, and more generally in connecting with the *divine feminine*.

Dowsing with willow rods for water

Willow has long been the dowser's friend, its supple, water-loving branches said to 'remember' the pull of underground streams. Cut fresh, the forked rod is shaped into a gentle Y, held loosely in both hands with the single stem pointing forward. As you walk, the wood is said to respond to the unseen flow below, twisting, tugging, sometimes snapping downward with surprising force when it crosses a vein of water.

Traditional water witches claim willow is more sensitive than hazel or ash, perhaps because of its own thirst for wet soil; it seeks out water as surely as its roots do. Old dowsers whispered that the rod does not merely detect; it *listens*, feeling the conversation between earth and water and translating it through the living wood.

Preparing the willow rod

Choose the right willow

Select a healthy, water-growing willow (*Salix alba, Salix fragilis*, or similar). Tradition favours willow cut from near running water, where its spirit is already attuned to flow.

Cut a forked branch in the shape of a Y, ideally during a waxing moon or in early morning when sap is rising. Folklore holds that freshly cut wood retains more 'sensitivity' to earth energies.

Shape and hold

Strip off any leaves or side twigs.

Hold the two forks lightly in your hands, palms up, with the single stem of the Y pointing forward and slightly upward. Your grip should be loose enough for the rod to twist or pull without resistance.

The act of dowsing

Ground yourself

Stand still for a few breaths, feeling your feet heavy against the earth. Some dowsers whisper to the willow or make an offering—a strand of hair, a few drops of water—to 'wake' the spirit of the wood.

Walk slowly

Move forward at a steady pace, letting your attention soften. Many dowsers describe this state as half-meditative, neither willing the rod to move nor forcing it still.

Reading the rod

When passing over an underground stream or a water vein, the rod may react sharply—twisting, dipping, or pulling downward, sometimes so strongly that the wood creaks or even snaps.

Some practitioners feel the rod 'drag' in the direction of flow, giving clues to both depth and movement of the water.

Marking the spot

Once a reaction occurs, pause, walk back over the area to confirm the pull, and mark the location.

Magical and spiritual considerations

Traditional water witches believed the rod does not merely sense water but listens to it, translating the conversation between the earth and underground streams into a physical language. The willow's own spirit, tied deeply to the lunar tides and water's secret pathways, acts as a mediator. Some practitioners prefer to work on moonlit nights, claiming that the rod responds more strongly when the moon's gravitational pull affects water tables. Others leave a small token, such as a berry or urine, to add to the water of the land, thanking the land for its gift.

Willow and the grieving process

During the 16th and 17th centuries, the association of willow became significant to the grief suffered by forsaken lovers, and they adopted the custom of wearing a cap or crown made of willow twigs and leaves. By the 19th century, illustrations of weeping willows often appeared as ornaments or *memento mori* on gravestones and mourning cards. This association is especially evident in the song 'All Around My Hat', which reflects this use.

The song 'All Around my Hat' is derived from a 19th-century English song. In an early version, dating from the 1820s, a Cockney costermonger vowed to be true to his fiancée, who had been sentenced to seven years' transportation to Australia for theft and to mourn his loss of her by wearing green willow sprigs in his hatband for 'a twelvemonth and a day', the willow being traditionally associated as a symbol of mourning. The song was made famous by Steeleye Span in 1975, whose rendition, it is said, may have been based on a more traditional version sung by John Langstaff.

Culpeper on willow

<u>Willow tree description</u>: *There are various kinds of the Willow-tree, but the most common is the White Willow, which I shall here describe. It grows to be a large tree. The bark is rough, and of a pale brown colour on the trunk and on the branches of a whitish grey. The leaves are long, narrow, sharp-pointed, and of a light green. The catkins are brownish. Botanists enumerate twenty-one kinds more, amongst which is the Creeping Salow, which scarcely rises to a foot high.*

<u>Place</u>: *It is common by the waters all over the land.*

<u>Time</u>: *The flowers appear early in spring.*

<u>Government and virtues</u>: *The Moon owns it.*

Both the leaves, bark, and the seed, are used to stanch bleeding of wounds, and at mouth and nose, spitting of blood, and other fluxes of blood in man or woman, and to stay vomiting, and provocation thereunto, if the decoction of them in wine be drank. It helps also to stay thin, hot, sharp, salt distillations from the head upon the lungs, causing a consumption. The leaves bruised with some pepper, and drank in wine, helps much the wind cholic. The leaves bruised and boiled in wine, and drank, stays the heat of lust in man or woman, and quite extinguishes it, if it be long used. The seed also is of the same effect. Water that is gathered from the Willow, when it flowers, the bark being slit, and a vessel fitting to receive it, is very good for redness and dimness of sight, or films that grow over the eyes, and stay the rheums that fall into them; to provoke urine, being stopped, if it be drank; to clear the face and skin from spots and discolouring. Galen says, the flowers have an admirable faculty in drying up humours, being a medicine without any sharpness or corrosion; you may boil them in white wine, and drink as much as you will, so you drink not yourself drunk. The bark works the same effect, if used in the same manner, and the tree hath always a bark upon it, though not always flowers; the burnt ashes of the bark being mixed with vinegar, takes away warts, corns, and superfluous flesh, being applied to the place. The decoction of the leaves or bark in wine, takes away scurf and dandruff by washing the place with it. It is a fine cool tree, the boughs of which are very convenient to be placed in the chamber of one sick of a fever.

In the fifty-third volume of the Philosophical Transactions, page 195, we have an account given by Mr. Stone, of the great efficacy of the bark of this tree, in the cure of intermitting fevers. He gathered the bark in summer, when it was full of sap, and having dried it by a gentle heat, gave a drachm of it in powder every four hours between the fits.

While the Peruvian bark remained at its usual moderate price, it was hardly worth while to seek for a substitute, but since the consumption of that article is become nearly as equal to the supply of it, from South America, we must expect to find it dearer, and very much adulterated every year, and consequently the white Willow bark is likely to become an object worthy the attention of the faculty; and should its success, upon a more enlarged scale of practice, prove equal to Mr. Stone's experiments, the world will be much indebted to that gentleman for his communication.

Charubel's commentary

Charubel has no commentary on the willow in his *Grimoire Sympathia*.

Personal experience

There are two types of willow in walking distance from my home. One is the white willow, and the other is the weeping willow. I have not worked with weeping willow as yet, but that is a tree I would like to work with at some point.

White willow is prevalent around my house and the second most populous plant after the alder. She is silver-green and has leaves and bark that are tough and grey-silver.

She came to me as a teacher; she knows of the earth's medicines and is a great plant to just sit and learn from. She talked to me about the horsetail that grows at her feet and about how old that plant was, a plant that could help me with my eczema and that I should talk to it.

She speaks in musical tones, like a bird, and in colours, in blues and greens that flow across the inner eye.

She is magnetic and emotional, showing me places in my body where I had blocked my feelings and stored them up for future use. She is of the moon, and I would place her as the moon in Gemini because she communicates freely and of essential things.

Personal meditation on the willow

When she comes, she comes with the breeze behind her. Softly she sits on the tree trunk, book in hand, a book I never quite get to see. Her voice is softly singing, melodious, but it seems to come through as if many notes are being played simultaneously. Music slides from tone into colours, blues and greens swimming around behind my closed eyes.

She tells me that I must concentrate on understanding; you must listen with the intent to learn. You must hone the blade of your mind to catch the drops of knowledge.

She softly speaks the secrets of the green world, advising whom to talk to next, who is a friend and ally, and who is not. She is soft in her demeanour, never reproachful, and smiles easily at each small step you take. Holding my hand when I stumble or mishear.

Understanding is just that, standing with the earth under your feet, listening for all you are worth for every word she says. Hoping it never stops as you bask in the flow of the words. They flow into you like water filling you up until bursting the banks of your mind. Forcing you to open to new ideas in gentle pushes until you succumb like the banks of a river crumbling away but with no feeling of loss. And then she closes the book and smiles, moving away.

It's time to go.

Learning is in the listening. Be still

Willow

Slowly and quietly, she comes
Bringing the breeze and the birdsong
Singing the knowledge into you
Moving the colours to open your mind
Pushing, ever pushing that bit further, that bit more
Till you spill over and your mind opens in flight
She talks to you of medicines
Of friends and foes to meet
She softly moves to you
And you sit at her feet
The pages of her book turn
And you desire for a look
But she is giving you the answers
You'll not find them in a book
So listen with your ears and listen with your mind
Her music flows inside you
It is gentle, and it's kind
Her teachings stay within you
Leaving ignorance behind

CHAPTER SIXTEEN

The Hazel

Hazel

The Celts equated hazelnuts with concentrated wisdom and poetic inspiration, as is suggested by the similarity between the Gaelic word for these nuts, *cno*, and the word for wisdom, *cnocach*.

There are several variations of the following ancient tale: nine hazel trees grew around a sacred pool, dropping nuts into the water, which were eaten by salmon that absorbed the wisdom. The number of bright spots on the skin of salmon showed how many they had eaten and thus how wise they were.

From an Irish variation of the tale, one salmon received all of these magical nuts. A Druid master, in his bid to become all-knowing, caught the salmon and instructed his pupil to cook the fish but not to eat any of it. However, in the process, hot juice from the cooking fish spattered onto the apprentice's thumb, which he instinctively thrust into his mouth to cool, thereby imbibing the fish's wisdom.

This lad was called Fionn Mac Cumhail, who went on to become one of the most heroic leaders in the Irish tradition. The story is very much like the story of Taliesin's creation, where he also burns his thumb on the cauldron of Cerridwen and therefrom absorbs the wisdom that the Goddess had so carefully prepared for her son.

Hazel trees frequently grow as a clump of slender trunks. They readily respond to coppicing, a practice which can extend and even double the lifespan of a hazel. Either way, people have put the young shoots or whips, as well as the thin trunks, to a variety of uses. It was once a tradition to give brides hazelnuts on their wedding day to pass on wisdom, fertility, and good fortune.[61] In Devon, this job fell to an old woman who waited outside the church to greet the bride with a basket of hazelnuts.[62]

Magical use of hazel

Hazel has long been a favourite wood from which to make staffs, whether for ritual Druidic use, medieval self-defence, as staffs favoured by pilgrims, or to make shepherd's crooks and everyday-type walking sticks. Hazel is also used in creating staffs for Morris dancers used in their traditional dances.

---

[61] Cf. S. Theresa Dietz, *The Complete Language of Flowers: A Definitive and Illustrated History* (New York: Wellfleet Press, 2020).
[62] Margaret Baker, *Discovering the Folklore of Plants*, 3rd edn (Boxley: Shire Classics, 2011 [1969]).

Forked twigs of hazel were also favoured by diviners, especially for finding water. Hazel leaves are usually the earliest native ones to appear in spring and often the last to fall in autumn, and were fed to cattle as fodder. It was also believed that they could increase a cow's milk yield.

Medicinal use of hazel

The bark, leaves, catkins, and fruits are sometimes used medicinally. They are astringent, diaphoretic, febrifuge, nutritive, and odontalgic (i.e. for toothache). The seed is stomachic and tonic. The oil has a very gentle but constant and effective action in cases of infection with threadworm or pinworm in babies and young children.

Hazel can be used to make a flour for bread or biscuits and is now is used as a major constituent in *nut milks*.

Modern research on hazel

Science is now confirming that hazelnuts bring a wealth of benefits to the brain, as highlighted by Gorji and colleagues. Their study reveals that hazelnuts provide neuroprotective effects, making them an excellent choice for preventing cognitive decline and neurodegenerative diseases. Packed with antioxidants, monounsaturated fats, and essential nutrients like vitamin E, hazelnuts help reduce inflammation, shield brain cells from damage, and enhance overall cognitive function. Regular consumption supports brain health by strengthening neural pathways, enhancing mental clarity, reducing oxidative stress caused by free radicals, and promoting neuroplasticity, which encourages the brain's adaptability and resilience.[63]

This aspect of neuroplasticity is particularly fascinating to me, as it resonates deeply with the spiritual exchange I had with hazel about the nature of change and the importance of adaptability. These findings not only validate hazelnuts' historical association with wisdom and communication but also establish their modern role as a powerful food for the mind.

---

[63] Narjes Gorji and others 'Almond, Hazelnut and Walnut, Three Nuts for Neuroprotection in Alzheimer's Disease: A Neuropharmacological Review of Their Bioactive Constituents', *Pharmacological Research*, 129 (2018), 115–27.

Culpeper on hazel

Hazel Nuts are so well known to every body, that they need no description.
Government and virtues: *They are under the dominion of Mercury. The parted kernels made into an electuary, or the milk drawn from the kernels with mead or honeyed water, is very good to help an old cough; and being parched, and a little pepper put to them and drank, digests the distillations of rheum from the head. The dried husks and shells, to the weight of two drams, taken in red wine, stays lasks and women's courses, and so doth the red skin that covers the kernels, which is more effectual to stay women's courses.*

And if this be true, as it is, then why should the vulgar so familiarly affirm, that eating nuts causes shortness of breath, then which nothing is falser? For, how can that which strengthens the lungs, cause shortness of breath? I confess, the opinion is far older than I am; I knew tradition was a friend to error before, but never that he was the father of slander. Or are men's tongues so given to slander one another, that they must slander Nuts too, to keep their tongues in use? If any part of the Hazel Nut be stopping, it is the husks and shells, and no one is so mad as to eat them unless physically; and the red skin which covers the kernel, you may easily pull off. And so thus have I made an apology for Nuts, which cannot speak for themselves.

Charubel's commentary

Again, there is no information on hazel in Charubel's *Grimoire Sympathia*.

Personal experiences with hazel

Hazel is one of my favourite trees, and it gave me my first staff, which I still have in my temple at the back of my house. The hazel provides food for the noisy and somewhat aggressive cohort of squirrels that live in my wood. There is a joke in my house that one day a plump squirrel who is balancing on the edge of a spindly hazel branch will bring down the entire wood with his weight. He never does—the hazel just bends and lets him carry on with his daily business.

This is what hazel teaches: the 'wisdom of bending with the winds'. It provides a lesson on how to embrace change, absorb it, resetting one's course, and moving on.

They come as a gender-fluid spirit, neither male nor female, but shifting between the two. Young and quick. Indeed, very chatty. They talk using a clicking tone, like branches rubbing together. I find it hard to listen to for long periods as their communication is extremely intense and very quick.

To me, the tree embodies the Greek Hermes, as displayed in his swiftness, his lithe intelligence, and his love of storytelling, as well as his mischievous nature, making him a bit of a trickster.

They teach me to move with the times, embrace change, and to roll with the punches. They teach me that all people in my life are visitors. The swift coming and going can teach you or indeed can break you. But how you allow it to impact your path is ultimately up to you.

You create your own environment through the actions of your mind and soul. It is your decisions that impact your world; you create reality, but sometimes you allow others to influence what you think.

You must be fluid but not fickle; hold on to your own truths because they are not worth any less than someone else's opinions. It is your perspective on life that is valuable because it is unique to you and not a reflection of someone else's. For we all contribute to that mercurial pool of 'memory', by creating our shared reality.

Personal meditation on hazel

When they come, it's loud, shrieking with laughter as they run full tilt into the wood. Breathless and sunny, they flop down at my feet. Broad-brimmed hat obscuring a young face. Then it turns upward to look at me. No covert communication here; this is full-on in-your-face conversation. Cheer up, buttercup, they say with a grin as big as the moon. Toothy and with a naughtiness that speaks of mischief. Why are you still, they ask, when motion is what is natural? Why do you sit when you could run?

Why do you worry and create realities in your mind that never come to be, but your pre-emptive mourning takes the energy you use to run, to play, and to experience the real rather than the imagined? When things change, you change. Move your ideas to make the new reality your own. Change is inevitable as doors open and friends leave; others fill the hole. Like mercury in the movement, never staying still, always moving, filling gaps you didn't know existed.

Accept change and change the world

Hazel

Here they come
All joy and gap-toothed smile
Faster than the speed of mischief
Life is a tapestry you must weave
Change is the Warp and the Weft
You are the weaver and not the thread
Change your ideas,
your faith,
your belief
Give change a chance and move the world
One change at a time
Prove your worth
Be plasticine
And sculpt new ideas
Be open to shifting your goalposts
Moving your mind
To new ways of thinking
Unfetter
Unlearn
Redefine

CHAPTER SEVENTEEN

The Ivy

The mythology of ivy

Ivy winds through human myth as persistently as it clings to stone, a plant of binding, devotion, and eternal life. Its evergreen leaves, untouched by winter's decay, made it a symbol of immortality and fidelity in both ancient and magical traditions.

In the ancient Celtic world, ivy was revered as sacred, representing the spiral of life, symbolising the soul's journey and the interconnectedness of all things. The way it spirals upward, twisting and weaving around tree and rock, mirrors the soul's ascent towards wisdom, though always by the winding, indirect path. Ivy's clinging nature was not seen as parasitic but as a lesson in persistence and in finding strength by holding on to others, making it a plant of friendship, loyalty, and binding oaths. In Druidic thought, ivy was associated with resonance and integration, teaching that all living things are interconnected by the threads of the same great pattern.

The Greeks and Romans wove ivy into the myths of Dionysus (also known as Bacchus), the god of ecstasy, wine, and divine madness. Ivy garlands crowned his priests and initiates, for the plant was believed to protect against drunkenness even as it celebrated wild abandon. Dionysus himself was sometimes depicted crowned in ivy instead of grapevine, for ivy was the ever-living, death-defying counterpart to the seasonal vine, hinting at the god's power over life, death, and rebirth. Ivy was also laid upon the tombs of the dead, a promise of resurrection and eternal memory.

In British folklore, ivy was one half of the sacred holly and ivy pairing, representing the feminine to holly's masculine. Ivy, as the softer, binding plant, was associated with women, intuition, and the hidden mysteries of the hearth, while holly stood for the hard, protective strength of the hunter and warrior. Old carols still echo this symbolic pairing.

Ivy's mythic lesson is always one of connection—to climb, it must cling; to survive, it weaves itself into others. In magic, it has long been used for binding spells, love charms, and protection, but also for deeper spiritual resonance: to remind us that no soul stands alone, that life spirals ever upward when we are entwined with the living web of all things.

Ivy in Celtic tree lore and Ogham

In the Ogham alphabet, ivy is represented by Gort, the twelfth character,[64] symbolising tenacity, binding, and spiritual growth through challenge. Just as ivy winds its way around other trees to reach the light, Gort teaches

---

[64] Yuri Leitch, *Ogham Grove* (Glastonbury: Yuri Leitch, n.d.).

that progress in life and magic often comes through persistence and the willingness to weave yourself into the greater whole. It is a reminder of both the blessings and dangers of connection: ivy can strengthen by holding on, but it can also smother if binding becomes control. Druids viewed it as a plant of soul resonance, used in meditation to attune to the great Web of Life, and in ritual to bind intentions into the spiral of fate.

Magical uses of ivy

Ivy has long been used to bind promises, especially in handfasting rituals, where ivy garlands symbolise eternal fidelity.

It is also worn as a circlet or hung above doors. Ivy is said to guard against negativity and psychic attack.

Two pieces of ivy bound together with red thread were traditionally carried as a charm to maintain loyalty between married people or lovers.

Sleeping with a sprig of ivy under the pillow was said to bring prophetic dreams and reveal a future lover's face.

Ivy's evergreen nature and spiral growth made it a plant of Otherworld journeys; it was sometimes woven into wreaths worn during Samhain rites to honour ancestors.

The darker uses of ivy

Ivy's spiralling, binding energy is not only a blessing—it can be turned to entrapment, control, and the slow crushing of another's freedom. Folklore warns of ivy's 'smothering embrace', where its persistence becomes domination, and this same quality has been used magically in curses and restrictive workings.

A poppet or carved image of the target is wrapped tightly in fresh ivy, each coil enchanted with words of binding, to trap a person in their own habits, lies, or guilt. Left to wither, the ivy symbolically tightens, restricting the target's will and sapping their vitality.[65]

Where ivy is used for fidelity, it can also be reversed. Binding ivy around a pair of intertwined candles and allowing them to suffocate in their own wax was a folk spell to break apart lovers by creating emotional entanglement and resentment.

---

[65] D. Valiente, *An ABC of Witchcraft Past and Present* (London: Hale, 1989).

Practitioners seeking to silence gossip or stop harmful speech would wrap ivy leaves around a stone, bind it with black thread, and bury it under the hearth or threshold of the offender's home, 'rooting' their tongue in silence.

In some traditions, ivy was used in baneful ancestor work; a wreath of ivy placed on a cursed grave was believed to summon the dead to 'hold fast' to the spirit of the wrongdoer, drawing them into sickness or misfortune until amends were made.

These practices carry warnings: ivy's energy, once set in motion, is persistent and can become difficult to reverse. Its nature is to *cling*—so it may just as easily bind the caster if not properly released. Many witches, aware of ivy's duality, insist on formally cutting ties with it after any binding or cursing work, offering water or a strand of hair in appeasement when unbinding the spell.

Ivy as a herbal medicine

Though less commonly used today, ivy (*Hedera helix*) has a long tradition in European herbal medicine, valued for its expectorant and anti-inflammatory properties. Ancient Greek physicians prescribed ivy leaves steeped in wine to ease respiratory conditions and to counteract the effects of alcohol—perhaps another reason for its close association with Dionysus.

Modern herbal practice recognises ivy leaf extract for treating chronic bronchitis, asthma, and chesty coughs, as it helps loosen mucus and soothe inflamed bronchial passages. Ivy contains saponins (notably hederacoside C), which increase bronchial secretion and relax smooth muscles in the respiratory tract. Preparations are typically in the form of standardised syrups or tinctures rather than raw leaves, as fresh ivy is toxic if ingested in quantity and may cause gastrointestinal upset or skin irritation in sensitive individuals.

Topically, ivy leaves were historically bruised and applied as poultices to reduce swelling or soothe arthritic joints; however, modern use for this purpose is rare owing to the risk of contact dermatitis. Internally, only controlled extracts should ever be used. Its historical use as a cleanser of 'bad humours' hints at its mildly stimulating and detoxifying qualities.

Modern research

Modern clinical research strongly supports the use of standardised ivy leaf extracts in respiratory medicine, particularly for bronchitis and persistent coughs. Randomised controlled trials demonstrate that the standardised extract EA 575 (marketed as Prospan) is both safe and effective for treating acute respiratory infections. A 2024 three-arm randomised trial showed that EA 575 was non-inferior and, in some measures, superior to combination herbal syrups in alleviating bronchitis symptoms over a seven-day period.[66]

A meta-analysis of placebo-controlled trials confirmed these findings, showing a rapid reduction in cough severity, with significant improvements noted after just two days. By the end of treatment, 18% of patients treated with ivy extract were cough-free, compared to 9% in the placebo group.[67]

Observational studies also suggest ivy may help reduce antibiotic overuse. A 2025 retrospective analysis of German paediatric cases reported that children prescribed EA 575 for respiratory illnesses were less likely to require antibiotics later in the year, and many returned for repeat prescriptions, indicating both efficacy and tolerability.[68]

Culpeper on ivy

It is so well known to every child, almost, to grow in woods upon the trees, and upon the stone walls of churches, houses, &c. and sometimes to grow alone of itself, though but seldom.

<u>Time.</u> It flowers not until July, and the berries are not ripe till Christmas, when they have felt Winter frosts.

<u>Government and virtues.</u> It is under the dominion of Saturn. A pugil of the flowers, which may be about a dram, (saith Dioscorides) drank twice a day in red wine, helps the lask and bloody flux. It is an enemy to the nerves and sinews, being much taken inwardly, but very helpful to them, being outwardly applied. Pliny saith, the yellow berries are good against the jaundice; and taken before one be set to drink hard, preserves from drunkenness, and helps those that spit blood; and that the white berries, being taken inwardly, or applied outwardly, kill the worms in the belly.

---

[66] Peter Kardos and others, 'Efficacy and Safety of a Single Ivy Extract Versus Two Herbal Extract Combinations in Patients with Acute Bronchitis: A Multi-Center Randomized, Open-Label Clinical Trial', *Pharmaceuticals* 18(5) (2024), 754.

[67] Felix Holzinger and Jean-François Chenot, 'Systematic Review of Clinical Trials Assessing the Effectiveness of Ivy Leaf (*Hedera helix*) for Acute Upper Respiratory Tract Infections', *Evidence-Based Complementary and Alternative Medicine* (2011), 382789.

[68] Andreas Völp and others, 'Ivy Leaves Extract EA 575 in the Treatment of Cough During Acute Respiratory Tract Infections: Meta-Analysis of Double-Blind, Randomized, Placebo-Controlled Trials', *Scientific Reports*, 12(1) (2022), 20041.

The berries are a singular remedy to prevent the plague, as also to free them from it that have got it, by drinking the berries thereof made into a powder, for two or three days together. They being taken in wine, do certainly help to break the stone, provoke urine, and women's courses. The fresh leaves of Ivy, boiled in vinegar, and applied warm to the sides of those that are troubled with the spleen, ache, or stitch in the sides, do give much ease. The same applied with some Rosewater, and oil of Roses, to the temples and forehead, eases the head-ache, though it be of long continuance. The fresh leaves boiled in wine, and old filthy ulcers hard to be cured washed therewith, do wonderfully help to cleanse them. It also quickly heals green wounds, and is effectual to heal all burnings and scaldings, and all kinds of exulcerations coming thereby, or by salt phlegm or humours in other parts of the body.

The juice of the berries or leaves snuffed up into the nose, purges the head and brain of thin rheum that makes defluxions into the eyes and nose, and curing the ulcers and stench therein; the same dropped into the ears helps the old and running sores of them; those that are troubled with the spleen shall find much ease by continual drinking out of a cup made of Ivy, so as the drink may stand some small time therein before it be drank. Cato saith, That wine put into such a cup, will soak through it, by reason of the antipathy that is between them.

There seems to be a very great antipathy between wine and Ivy; for if one hath got a surfeit by drinking of wine, his speediest cure is to drink a draught of the same wine wherein a handful of Ivy leaves, being first bruised, have been boiled.

Charubel's commentary on ivy

The botanical name for this family is Aralicece. It is possibly the best-known member of the vegetable Kingdom. It may be found everywhere, even in parts of our large towns and crowded cities, attached to fretted walls of antique dwellings. The Ivy belongs to our much cherished and dearly beloved evergreens. It will compare favourably in loveliness with more exalted species, under favourable conditions, where its growth is profuse. It is affirmed by botanists that there are about 150 species of Ivy scattered over the earth. Still, it is with the common Ivy that I have to deal in the present article, as it's the plant that contains the most significant number of occult properties.

The Ivy held a prominent place among the ancient Greeks on festive occasions, especially during the festival of Bacchus, who was designated the God of wine and the son of Jupiter. Jupiter, rightly interpreted, signifies a young father, notwithstanding that he had a lewd and drunken son in the form of Bacchus. It is even so among the denizens of earth, many a good father has a drunken son. The feast of Bacchus was celebrated at the altar when the grapes had been gathered and during the vintage.

It was on these occasions that rejoicings of the most boisterous character were exhibited and when the excesses most degrading were indulged in amid the shouts of the mad and drunken crowd. It was on such occasions that crowns were worn on the heads of these drunken worshippers. These crowns were composed of Ivy intertwined with vine leaves.

There is a hidden or occult meaning in this combination. On the one hand, the fruit of the vine, which was caught, was the cause of that mirth and hilarity, symbolised by the vine leaves. On the other hand, the Ivy symbolises sobriety, durability, wisdom, and immortality.

The old astrologers assigned Ivy to the planet Jupiter, and I will therefore call this plant another centre of Jupiter. Since this is the case, the Bacchanal crown represents the two sons of Jupiter.

Hence, the Vine and Ivy represent two brothers of very different characters it must be confessed. The vine represents hilarity and mirth, which panders to the lower delights. The Ivy, on the other hand, symbolises wisdom and higher aspirations.

The Ivy may be found on any old hedge cop in rural districts. Where there is no tree adjoining to which it may cling to with its tenacious holdfasts it would redesign itself in its lonely bed and become a Bush having fully developed leaves with both bloom and Berry. I have witnessed developments in shady places and congenial spots in Wales.

But should there be a tree within reach it will make use of it in order to raise itself to get to pure air and also that it may partake of a little more light which the trees of the field enjoy. The Ivy appears to be conscious so to speak of its own weakness, it cannot stand erect without support, it cannot raise itself by its own stem. But when it comes into contact with the tree, it will choose to love it. It is not particular as to the object of its choice. It matters not what kind of tree it may be, whether a dumpy tree or a stately one, whether it is a big tree or a small one, whether it is rough or smooth; it is all the same. Anyone will do for its selfish purpose.

The tree, in its turn, does not appear to appreciate the embraces of the Ivy. I fancied it to say "oh, you wretched crawler, you lowlife thing, how you hurt me, how you disfigure my stately trunk. If you keep all, you will hide me altogether from the admiring gaze of my numerous friends."

That's what Ivy says: "I do not rob you of your sap. I supply my own wants from my own roots. I do, but I make use of your strength so I may gain a little more in the light and air and sunshine that your upper branches bask in and enjoy." And thus Ivy clambers up that sturdy trunk and envelops the stately stem with its never-relaxing grip.

The picture I present to you affords two important lessons. In one light, the image illustrates outright selfishness, highlighting those among the human race who, with their specious friendship, are worse than open enemies. They make use of the subject of their caresses merely they promote their own selfish ends. They have a sole motive while appearing to serve you. In reality, they elevate themselves at the expense of their confidential friends.

But there is another light in which this picture may be seen. The Ivy is a symbol of wisdom, immortality, and the higher life. As such, it teaches us to look on all sublunary things as subordinate to that higher life; hence, the Ivy will make use of even the noble oak in order that it may clamber up to its higher and purer life.

I have already suggested that the vine and the Ivy are brothers, and I may here offer an additional remark that there is a wonderful and striking likeness between them. The difference lies in this: the leaves of the vine perish annually, whereas the leaves of the Ivy are Evergreen or at least perennial, as there is no one leaf on any tree that retains its leaves for use, much less forever. Each leaf folds in its turn but it leaves a successor who reigns in his stead.

<u>Ivy on the soul plane</u>

I am a little doubtful as to whether you do, after all, retain anything like a true conception as to what it is really implied by the phrase soul plane. Now, when I see a plant on the soul plane, it must be understood that what I see from my present standpoint is an ideal, an ideal existence. It is in this light I desire these descriptions to be understood.

To further explain myself: the whole of the visible universe exists in its true form, but as a shadow. The real substance is what mankind are in the habit of designating the unseen.

An apostle once made the remark that, in conjunction with others of like faith, he looks not at things as they are seen, but at things as they are not seen. For the things that are seen are temporal, but the things that are not seen are eternal.

The unseen are the ideals, and these are imperishable. The ideals are our projections from the infinite mind, primarily considered; secondary, but also considered, are the ideas of those countless millions of agencies that work by the eternal and immutable laws. To use but a meagre illustration, the architect had a plan and a specification in his office before a stone of the palace was laid one on top of the other. There is not an item of superstructure which had not a prior existence in the mind, but should that building perish by storm, by flood or by fire, the ideal would prove indestructible.

This ideal universe has been projected by the grand architect on the spacious canvas that we call the soul world. Every idea is there, the past is there, the future is foreshadowed, and the prophet, who may be divinely inspired, can see on that very table the imperative, the imperishable archetype.

It is along the same lines that every plant exists forever in the ideal, and it is by virtue of this celestial existence that the terrestrial plant leaves and blooms on earth.

The Ivy, as perceived by me on the soul plane, is a plant that much resembles its earthly counterpart, with this difference. The psychic plant raises itself spirally, each layer resting on top of the other, ascending to a height of what appears to be 10 to 12 feet. It has the semblance of a tree trunk, with a bushy top covered in a white bloom the size of a rose.

The Infirmities for which Ivy is specially and pathologically applicable are the following. First, that of an overexcited brain, the subject of hallucination. Secondly, sleeplessness from overexcitement, and thirdly, its moral influence, which begets patience and resolution.

Personal experience with ivy

I have to disagree with Charubel on ivy's planetary rulership. He places her under Jupiter, but to me there is nothing jovial or expansive about her. Culpeper names her rightly as a Saturnian plant, and I would go further, ivy is Saturn in Cancer. There is discipline and gravity in her, yes. Still, it is the discipline of care, of a mother watching over her children, not the cold and distant Saturn that binds without love. She is stern because she loves, protective because she must be. And when I worked with her, there was also a dreaming, watery quality, the pull of fantasy and vision, the way she opens the mind to dream and memory. This is no airy Jupiterian generosity; this is the slow, tidal wisdom of Saturn working through the watery shell of Cancer.

Ivy does not come to me as a meek clinger or a smothering vine, but as a dragon, emerald-scaled, sinuous, twining around me with a strength that is firm yet gentle. Her coils are not constrictive; they brace me, as though she is holding me steady against the weight of the world rather than binding me to it. Her presence feels like a heartbeat pressed to my own, a deep, ancient rhythm that speaks of the green of the land and the long promise of the soil.

It amazes me now that I never saw the dragon in her before working with her. It seems so obvious. Look at her leaves—their pointed, proud tongues, each one shaped like the flick of a dragon's tongue. She has been showing herself all along, whispering her nature, and I walked past her for years thinking her only a creeping plant. Now I see her as she truly is: a guardian of the Green, ancient, patient, always watching.

When she winds around me in vision, she shows me the world through her green eyes—the tapestry of moss and roots, the slow persistence of stone cracking under roots, the breath of the earth itself. Her message is that if the world is to live, we must protect the Green. We must support one another as humans on the planet, including all individuals, regardless of our differences.

Then her tone sharpens. She shows me the stagnation of the air, the heaviness of exhaust fumes, the sky dulled by factory smoke, the slow choking of the world in the name of convenience. She presses the truth into me like a scar; we are trading the breath of the earth for speed, plastic, and ease.

Her coils tighten around my chest, firm but not cruel, and I feel what she wants me to feel: the struggle for breath. In that moment, I see children, wheezing, their skin blotched with rashes, their small lungs fighting for each shallow gasp. Ivy does not need to say it, but I hear it clearly: **this is our doing**. Asthma, allergies, the strange fragility of our children—these are not curses cast by fate, but the direct cost of polluted air, stripped soil, and the Green we have neglected to defend.

Ivy, as dragon, is no soft spirit. She is protective, yes, but her protection is fierce, demanding. She does not simply hold me; she steadies me and demands that I stand. Her final words are not words at all but a great exhale, a rush of leaves in storm-wind, carrying her truth into me:

The Green is the breath of the world. Protect it, and you protect the children. Forget it, and you forget the future.

A personal meditation with ivy

I go to her. She waits in my wood, unwinding herself from the oak, sliding down its trunk in slow, deliberate spirals until she curls around me. Her body is cool against mine, and her coils are not cruel—they are firm, supportive, like being held by something that understands both strength and gentleness.

Her voice fills the air between us. It is half the purring of a cat, half the cracking of wood just before the tree falls, a sound that is both comforting and quite unsettling, carrying the weight of truths I might not want to hear.

We talk, as we always do, of support and holding. She tells me that to hold someone is not to smother them, not to bind them so tightly they cannot grow, but to brace them, to be the trunk against which they can lean when the wind blows too hard. And yet, she says, protection does not mean shielding from all pain. Sometimes, you must let those you love experience the thing itself, even when it hurts, because only through that experience can strength be grown. Her coils tighten slightly as she says this, as if to remind me what it feels like to be held—not caged, but supported.

Then she speaks of the air. Her tone grows harsher, the crack of wood louder than the purr.

"The air is poisoned," she tells me. *"You know this. The children cannot breathe, and still you go on."*

Images fill my mind—the tight chests of little ones, the rasp of each breath, the red-rimmed eyes of those who grow up beneath skies they can no longer trust.

"You make medicines each day," she whispers, *"and you sell them. New bottles, new cures, new promises. But if you truly wished to heal, you would stop. You would give up your convenience. You would clear the air, and you would never need these cures again."*

Her coils tighten around my chest, pressing until I cannot draw a full breath. The sensation is uncomfortable, almost frightening, and she does not loosen until I feel panic stir.

"This," she growls softly, *"is what it is to live without clean air. This is how it feels for them—the children. This is what you choose for them every time you choose convenience over the Green."*

When she releases me, I gulp the air as if I have been drowning. The purr returns, softer now, curling around me like leaves in a breeze.

"Protect the Green," she says, her voice settling back into that rhythm of purr and crack, *"and you protect the breath of the world. Forget it, and you will learn to live without air—just as the children are learning now."*

She does not leave me. She never does. She simply unwinds again, returning to the oak, patient and eternal as she waits for me to listen.

Forget the Green and forget the future

Ivy

She coils from the oak,
a green dragon twining,
Scales glinting with the sorrow of the land.
Her touch is firm, not cruel—
A brace against the breaking of the world.
She tightens,
And I feel the weight of poisoned skies,
Lungs straining like sap trapped in frost.
Children's wheezing ghosts ride the wind,
Their breath stolen by our hunger for ease.
When she unwinds,
The air tastes like a promise I do not deserve.
She slides back into shadow,
But her grip lingers,
And my chest still hums with the truth
The Green is the breath of the world,
And we are choking it for comfort.

CHAPTER EIGHTEEN

The Apple

The mythology of the apple tree

The apple tree has always felt to me like a quiet threshold, a step in between the state of knowing and unknowing. Sit beneath an apple tree long enough, and you will feel it watching, with the deep, knowing kindness of something that has stood between the human and the Other for longer than we can remember.

I learned this truth for myself in the orchard of the Red Spring in Glastonbury, many years ago, before I had started working with plant or tree spirits. Up at the top of the orchard stands an ancient apple tree, and one morning, I sat beneath him for a long time in a sort of mesmerised state. I did not speak. I could not. I sat in floods of tears, not from sorrow, but from the sheer beauty of the energy that moved through that orchard. There was a presence there, ancient, patient, and utterly kind. The air felt alive with a low hum, as if the orchard itself were breathing around me. That tree, gnarled and heavy with age, felt like a teacher, and in that moment, that this tree was someone really special, I think of him often. I hope he is still there, standing sentinel at the top of the orchard, waiting. The next time I visit Glastonbury, I will stop and leave him a gift, a small act of gratitude for the first lesson he ever taught me.

Across stories, the apple has always been a tree of crossing over, of tasting what should perhaps not be tasted, of learning truths we are not quite ready to learn.

The Celts knew this well. The apple is tied to the Ogham letter *Quert*, a symbol of choice, beauty, and the delicate balance between life and death. In the old tales, a single apple branch, heavy with silver blossoms and fruit, could open the way to the Otherworld. The Isle of Avalon, the *Isle of Apples*, was a place of healing and eternal youth; priestesses ruled there, and to taste its fruit was to taste immortality. Apples were gifts from that realm, sometimes given in love, sometimes as a lure—because to bite into an apple from the Otherworld was to bind yourself to it.

The Norse spoke of Iðunn's apples, shining with youth, without which even gods wither. And in the Greek tales, golden apples belonged to the Hesperides, guarded by serpent and nymphs, gifts of immortality and symbols of desire.

But it is in traditional witchcraft that the apple takes on one of its most subversive and powerful meanings. It is the gift of sentience itself. In this telling, Lucifer, the Lightbringer—the true bringer of knowledge, not the demon of Christian fear—took the shape of the serpent and offered Eve the apple in Eden. This was not temptation in the way the priests like to tell it; it was an act of love and liberation. Lucifer gifted humanity with awareness, consciousness, the ability to choose, learn, and grow. It was the first initiation.

I often wonder if women have always been the conduits of this education. The old stories whisper that the *Nephilim*, those fallen Watchers, married mortal women and gifted them the secrets of astrology, the patterns of the stars, and the knowledge of plants and science. Were these women the first priestesses, the first witches? Is this where the quiet tradition of giving an apple to a teacher comes from, an echo of that first forbidden lesson, that first taste of illumination offered by the Lightbringer himself?

Even the Christians, perhaps unknowingly, honoured the apple as the fruit of knowledge, the bite that pulled us from innocence into awareness. The priests might have cursed it as sin, but those of us who walk the crooked path know better. Sometimes wisdom comes with a cost, and sometimes it is a cost worth paying. You cannot return to who you were before you tasted it, and perhaps you are not meant to

The magic of the apple tree

The apple is a tree of choice, healing, and thresholds. Its Ogham, *Quert*, asks you to stand at the crossroads of your own life and decide which branch you will follow. Apple magic is kind, but it is never soft. It will not let you hide from what you already know.

Wassailing the apple trees: a rite of reciprocity

In the heart of winter, when the earth sleeps beneath frost-crusted soil and the bones of the orchard lie bare, the people gather. With ribbons and cider, pans and poetry, they come not to take, but to give, to bless, to praise, to wake the sleeping spirits of the apple trees. This is *wassailing*, one of the oldest surviving rites of agricultural magic still practised in Britain today. Far from mere folklore, it is a ritual of reciprocity—a covenant between humans and trees, community and land.

At its heart lies a simple truth: that relationship births abundance. The trees are honoured as kin, not commodities. They are sung to, cajoled, and cheered; their trunks splashed with last year's cider, the first fruit returned in offering. Toast soaked in ale is placed in their branches for the robins—those tiny red-breasted messengers who carry prayers between worlds. And the people circle the trees, banging pots, firing off shotguns, or shouting to scare away malevolent spirits who might wither the harvest to come.

But the magic here is not in the noise or even the cider. It is in the mutual recognition—the understanding that the orchard gives not because it is made to, but because it is loved. Through this act of reverence, the trees are reminded that they are seen, valued, and sung to. And in turn, they thrive. It is a kind of enchantment made not through dominance but through dialogue.

Modern science is just beginning to catch up with what wassailers have always known: that plants respond to sound, to care, to intention. Mycorrhizal networks, tree hormones, pheromonal signals—these are the biochemical echoes of the same truth.

When we approach with song and gratitude, we don't just stir the spirit of the orchard; we awaken a web of living response. The trees know us, just as we know them. And when we give joy, they give fruit.

Wassailing, then, is not a superstition. It is a contract. It is the remembering that growth is not a one-way extraction, but a dance of giving and giving again. In this winter rite, we feed not just the trees but the unseen currents that run between land and human, magic and matter. We bless the orchard, and the orchard blesses us in return.

Protective and divinatory magic

Cut an apple crosswise, and you will find the five-pointed star of seeds, a pentacle placed there by nature herself. This is why apple wood makes powerful wands, charms, and protective talismans. The old games at Samhain, bobbing for apples, peeling them to divine lovers' names, were not just rural pastimes but echoes of older rites of love and fate. Apples have always been diviners of truth.

Otherworld work and spirit travel

The old Irish bards carried silver-apple branches into trances, their music guiding them to the Otherworld. At Samhain, an apple left on an altar or at a crossroads serves as an invitation to the spirits of the dead, offering a bridge between the living and the unseen.

Love and desire

The apple, sacred to Aphrodite, is still potent in love magic. Carve your name and another's into an apple, bind it with red thread, and bury it beneath a blossoming tree to strengthen a bond. But beware—apple magic reveals truth. It will not create love where none exists.

The healing wisdom of the apple—modern research and old magic

It is a curious and wonderful thing when the old magical truths whispering through folklore are confirmed by modern science. We have long known the apple as a tree of healing, a giver of vitality, and a gentle guardian of well-being. The Celts called it a tree of choice and balance, the Norse gave it to their gods as food of youth, and the old witches knew it as a charm for health and love. Today, research confirms what our ancestors already knew in their bones: the apple is medicine as much as it is myth.

Modern studies have shown that apples are rich in powerful polyphenols, including flavonols, procyanidins, and quercetin, which act as potent antioxidants and reduce inflammation throughout the body. These compounds protect against heart disease, asthma, and certain cancers by calming the body's inflammatory pathways. Cardiovascular health is one of the apple's strongest gifts. Clinical trials have shown that regular apple consumption can lower LDL cholesterol—sometimes by as much as 24% in postmenopausal women—and improve the elasticity of blood vessels, which helps reduce high blood pressure and protect against atherosclerosis.[69]

The apple's influence on metabolism is equally remarkable. Its polyphenols regulate blood sugar and improve insulin sensitivity, lowering the risk of type 2 diabetes. Observational studies consistently show that individuals who regularly consume apples have a significantly reduced risk of developing diabetes, and animal studies suggest that this is due to quercetin and phloridzin improving glucose control.[70] Apples are also rich in soluble fibre, particularly pectin, which acts as a prebiotic, nourishing beneficial gut bacteria and helping regulate digestion.

---

[69] Nicola P. Bondonno and others, 'The Cardiovascular Health Benefits of Apples: Whole Fruit Vs. Isolated Compounds', *Trends in Food Science & Technology*, 69 (2017), 243–56.
[70] Gayer and others 2022 [author to supply reference]

A healthy gut microbiome, in turn, supports weight management and lowers obesity-related metabolic risks.[71]

The apple's magic stretches further still. A growing body of evidence links apple consumption to lower risks of colorectal and breast cancers, with polyphenols influencing cell cycle regulation and slowing tumour growth.[72] There are even whispers in medical journals of its effects on the mind—studies suggest apple antioxidants protect the brain from oxidative stress, slowing cognitive decline and supporting long-term brain health.[73]

How wonderful that the old sayings carry such truth. The familiar phrase "an apple a day keeps the doctor away" is no empty rhyme. In fact, a 2013 modelling study published in the *British Medical Journal* suggested that eating an apple daily for people over 50 could lower cholesterol as effectively as a statin, but without the side effects.[74] Perhaps there is also something older and deeper in the habit of giving an apple to a teacher. We witches might well smile at this, remembering Lucifer, the Lightbringer, the serpent in Eden who gifted Eve the first true lesson of consciousness, and the Nephilim—those fallen teachers—who imparted the secrets of the stars to their mortal wives. The apple has long been associated with knowledge, healing, and transformation. Science, it seems, is simply catching up.

Culpeper on apple

Description. This is a tree so well known for its fruit that it would be needless to give any description of it here. Among the numerous varieties of apples, those that are accounted best for medicinal use are the pearmain and pippin, yielding a pleasant, vinous juice with a little sharpness.

---

[71] Tingting Jiang and others, 'Apple-Derived Pectin Modulates Gut Microbiota, Improves Gut Barrier Function, and Attenuates Metabolic Endotoxemia in Rats with Diet-Induced Obesity', *Nutrients*, 8(3) (2016), 126.

[72] Shufang Yang and others, 'Evaluation of Antioxidative and Antitumor Activities of Extracted Flavonoids from Pink Lady Apples in Human Colon and Breast Cancer Cell Lines', *Food & Function*, 6(12) (2015), 3789–98.

[73] Hao Yang and others, 'Apple Polyphenol Extract Ameliorates Atherosclerosis and Associated Cognitive Impairment through Alleviating Neuroinflammation by Weakening TLR4 Signaling and NLRP3 Inflammasome in High-Fat/Cholesterol Diet-Fed Ldlr−/−Male Mice', *Journal of Agricultural and Food Chemistry*, 71(42) (2023), 15506–21.

[74] A. Briggs and others, 'A Statin a Day Keeps the Doctor Away: Comparative Proverb Assessment Modelling Study', *BMJ* (2013), 347.

Place. It is well known to grow in orchards and gardens.

Time. Different kinds flower at different times; all between April and the latter end of May. The John apple, which is the latest, is not ripe till October.

Government and virtues. Apple-trees are all under the dominion of Venus. In general they are cold and windy, and the best are to be avoided, before they are thoroughly ripe; then to be roasted or scalded, and a little spice or warm seeds thrown on them, and then should only be eaten after or between meals, or for supper. They are very proper for hot and bilious stomachs, but not to the cold, moist, and flatulent. The more ripe ones eaten raw, move the belly a little; and unripe ones have the contrary effect. A poultice of roasted sweet apples, with powder of frankincense, removes pains of the side: and a poultice of the same apples boiled in plantain water to a pulp, then mixed with milk, and applied, take away fresh marks of gunpowder out of the skin. Boiled or roasted apples eaten with rose water and sugar, or with a little butter, is a pleasant cooling diet for feverish complaints. An infusion of sliced apples with their skins in boiling water, a crust of bread, some barley, and a little mace or all-spice, is a very proper cooling diet drink in fevers. Roasted apples are good for the asthmatic; either raw, roasted or boiled, are good for the consumptive, in inflammations of the breasts or lungs. Their syrup is a good cordial in faintings, palpitations, and melancholy: The pulp of boiled or rotten apples in a poultice, is good for inflamed eyes, either applied alone or with milk, or rose or fennel-waters. The pulp of five or six roasted apples, beaten up with a quart of water to lamb's wool, and the whole drank at night in an hour's space, speedily cures such as slip their water by drops, attended with heat and pain. Gerard observes, if it does not effectually remove the complaint the first night, it never yet failed the second. The sour provokes urine most; but the rough strengthens most the stomach and bowels.

Charubel's commentary on apple

There is no commentary on Apple in *Grimoire Sympatheia*.

Personal experience with the apple

The apple tree has always been a teacher to me, but not the kind who sits you down and pours out wisdom in great torrents. No, he is an old gardener, slow and careful, an adviser who tells me things in small pieces, like spoon-feeding a child something he knows is good for them, but isn't sure they're ready to swallow all at once.

When he comes to me, he is always the same: boots caked in earth, shirt creased from work, hands rough with soil, smelling of crushed leaves and damp wood. His presence is quiet but firm, and he has the kind of patient humour that makes you feel both comforted and slightly chastised. He looks at me with those knowing eyes, as if he can see every tangle of roots in my life, every wrong turn, every seed I have failed to water. But he does not scold. He simply tilts his head, folds his arms, and asks questions that dig deeper than any spade.

What will you learn from this choice? Are you looking for the easy way or the way that will make you grow?

That is his way: never a direct order, always a choice laid out before me like two paths in the orchard. One path soft with grass, the other rough and stony. And he always nudges me towards the stony one. Not because he is cruel, but because he knows that it is the hard ground that makes roots strong.

It was after a long meditation with him that I made one of the most difficult choices of my life. I had been miserable in my job for years, suffocating under a head teacher who treated the pastoral department—the department I was responsible for—as little more than an inconvenience. I sat with the old man beneath the apple tree, and he didn't tell me what to do. He just looked at me, waiting until I blurted out what I already knew.

So, he said at last, *will you stay and grow bitter, or will you take the choice that teaches you something, even if it's hard?*

It was as if he handed me a spade and left me to dig my own soil. The very next day, I confronted my head teacher, told him exactly how dissatisfied I was with the way he treated my department, and handed in my request for early retirement.

It was not an easy choice. It has not, in some ways, made my life easier—losing the security of a regular income is no small thing. But it has opened my life in ways I could not have imagined. My days are now brighter, fuller, filled with beauty and freedom, with writing, teaching, and walking the old paths more deeply than ever before.

The old gardener was right. The easy choice might have kept me safe, but it would not have made me grow.

So, if hard choices are coming your way, go and talk to Old Man Apple. Sit beneath his branches, breathe in the scent of leaves and ripening fruit, and listen. He will not give you comfort. But he will give you truth. And sometimes, that is the greater gift.

Personal meditation with the apple

I close my eyes and breathe, letting the world fall away. It doesn't take long for him to come; he never hurries, but he always comes.

The first thing I feel is his presence. It settles over me like the weight of damp earth, grounding, certain. Then I see him, just as he always appears: the Old Gardener. His boots are scuffed and caked with dried mud, his shirt is worn and faded, sleeves rolled to his elbows. His hands are brown with soil, rough and strong, nicked here and there by thorns or tools. There's a smear of crushed green on his forearm where he's wiped his hands, and he smells of grass, sap, and apples beginning to fall and soften in the orchard. His face is lined, weathered by years of sun and wind, and there's that half-smile of his, kind, but with the faintest edge of amusement, as if he already knows everything I'm about to say.

He doesn't speak straight away. He never does. He just stands there, folding his arms, one eyebrow raised, and I can feel him waiting. His silence makes me restless, because under his gaze, I feel like all my secrets have been tipped out onto his workbench, laid out like little seeds, and he is quietly sorting through them. He won't touch them, won't name them aloud, but he *knows*.

When he finally speaks, his voice is warm but solid, each word dropping into me like a spade turning soil.

So. You've come because you're at a choice again.

I swallow, because he's right. Of course, he's right. He tilts his head at me, that wry smile tugging at his mouth.

You already know which road you're going to take. But you're hoping I'll let you off the hook, give you an excuse. Is that really what you want?

The words land in my chest like a stone. I don't answer because there's no point. He knows.

He steps closer now, crouching so we are eye to eye. His eyes are dark, brown as deep, wet soil, and there's kindness in them, but no softness.

The easy road will keep you safe. But you won't learn anything on it. The other one—he gestures with one soil-streaked hand—*that one will be hard. It will break you a little. But it will teach you more than you think you want to know. That's the road worth taking.*

I feel it then, exactly as I always do. My stomach tightens, my chest aches, doubt pressing heavy against my ribs. I want to turn away, to argue, to ask why he can't just let me rest for once. But beneath the fear, something shifts. Slowly, like roots pushing down into dark earth, a new feeling spreads through me. Solid. Strong. It starts at my spine, moving outwards, anchoring me. The fear doesn't vanish, but it changes. It becomes weight, real, necessary, the weight of a seed just before it splits open.

He watches me quietly, head tilted, and then he gives me a slight nod as if I've passed some silent test.

There now, he says at last, voice low and certain. *You've already chosen. You just needed to feel it in your bones.*

And I realise he's right. I *have* chosen. It will not be an easy road, but it is the right one, the one that will stretch me, grow me, make me something more than I am now.

The Old Gardener straightens, dusting soil from his hands. He gives me that half-smile again, all fondness and exasperation in one.

Off you go, then. Hard road or not, there's work to do.

And just like that, he turns back to his tree, humming to himself, his tune low and steady, like bees murmuring in blossom. I know he will be there the next time I need him, waiting with the same patient eyes, ready to hand me another choice I might not want but will need.

Every choice is an initiation into who you truly are

Apple

To choose is freedom
He says in his way
To choose is hard
But you must have your say
The lessons you learn
They all point the way
To the person you are at the end of the day
So learn to choose hard
And learn to choose right
Even when might says move this way
And others fall short
Choose for yourself
And pay others no mind
Choose to be beautiful
Choose to be kind
And following choice
Ride the changes that come
And knowing that you are free
Turn your face to the Sun.

WALKING IN THE GREEN

So, here we are at the end of our time together, though there is no true end to this path, only a new beginning if you choose to take it. Books can only carry you so far. The rest of the work lies in your hands, in your heart, and in your willingness to live differently.

My time with the trees has changed me. They have stripped away my excuses, challenged my comfort, and taught me truths I would rather not have faced. And yet, in doing so, they gave me the courage to live more honestly, more gently, and more deeply in love with the world. The changes they asked of me were small but profound, and they have reshaped not only my magic but my entire life. That is how this great work moves: quietly, insistently, in small steps that ripple outward until they touch everything.

Now, I invite you to examine your own practice and life. Be honest with yourself. Where are you walking in alignment with the living world, and where are you still trapped in the habits of convenience, consumerism, and ego? Where are you still choosing the easy path rather than the right one? These are not questions to be asked once and forgotten—they must be lived, answered again and again in the choices you make each day.

The trees will help you if you let them. So will the mosses, the herbs that grow at your feet, the weeds that push through the cracks in the pavement. Make contact with the land around you, not in some abstract 'nature connection' way, but as you would with a friend. Speak to it. Learn its moods. Offer it your service—whether that's planting for pollinators, protecting wild places, or simply tending a single square foot of earth with love. Be in relationship with it.

And let your magic be part of that service. Stop keeping magic locked away as something separate, something only for ritual circles and formal workings. Life itself is the ritual. Every breath, every word, every choice is magic if you let it be. Bless your morning tea, whisper gratitude to the rain, and lay protection charms in your garden as easily as you wash the dishes. Magic is not a thing you do; it is a way you live.

But to live this way, you must also turn away from the illusion they would have you believe. Every day, our senses are assaulted by a continuous stream of fear news designed to make you anxious, hopeless, and small. Wars, disasters, endless screaming headlines—most of which you can do nothing about except wring your hands. Do not give them that power. Turn it off. Step out of the illusion. The world they present is not the whole truth. Your magic, your service, your daily choices—that is where real change happens.

Concentrate on your own life. Make a difference with the people close to you. Show, by the way you live, that there is another way to be. Because that is how change spreads—not by shouting at strangers online, but by living in such a way that others feel it, see it, and begin to wonder if there might be another path.

Show them that happiness does not live in the car you drive, the size of your house, or how many people you can make feel small to make yourself feel big. Show them that true happiness is quiet and rooted—it is satisfaction, contentment, the warm knowledge that you are living in right relationship with the world.

And the greatest act of rebellion against this consumerist sickness is simply this: define 'enough' for yourself. Refuse to let them tell you what it should be. Once you know your own 'enough', you are free—because you cannot be bought, and you cannot be controlled.

Remember this, above all: we have already won the greatest race simply by being here. Out of all the possibilities, all the countless chances, you were born into this world—this astonishing, breathing, sacred Earth. That alone is a miracle, and to live as if it is anything less is a betrayal of the gift we have been given.

So take this chance. Step into the work entirely. Be reborn into a new way of being—one rooted in service, in love, in kinship with our wild green brothers and sisters. Let the land change you, and then let that change ripple outward. There is still beauty here. There is still hope shimmering in the dew, still wonder beneath the golden sun, songs in the forest waiting for us to join in, not as masters, but as kin.

The time is now. Not tomorrow. Not 'when things settle down'. Now.

So walk out into the Green. Walk it. Live it. Be it. Let every step, every breath, every choice be a spell. And when the Green calls you—and it will, if you are listening—answer.

Walk in the Green, my friend. And do not look back.

Black Paths and Green Cathedrals

The trees, the green, the people
They rise above our heads
They speak with ancient voices,
in the movement of the leaves
That once were heard in whispers,
but now are heard in screams
As our ears are deaf to others, the not us,
the other things
But if we still our clamour, listen closely,
hear them sing
They make the air around us, in the circle where we meet
They are the great providers; of everything we eat
And yet we pay no homage, no gratitude or grace
Instead, the root and branch we rip,
with plastic we replace
We look for other Edens in the blackness out in space
And standing in the green light,
The people are unseen
Our spirits are so stunted,
so small and hard and mean
But when we stand in forests of oak and birch and thorn
We will remember who we were,
when we were first newborn
Our spirits rise to be there,
our hearts are not so worn
If we listen with minds open,
we will see where we belong
With feet upon the good black earth and eyes up to the green We can
see the Eden of the Earth,
the source of all our being
And then as Gaia's Glorious child,
wild-haired and newly Free
We will walk along the black paths
And in Cathedrals made of trees
And then at last will finally
find our spirits ease.

Sian Sibley, 2025.

BIBLIOGRAPHY

Adlof, Cassidy C., 'How Does Harvesting Impact White Sage (*Salvia apiana*) as a Cultural Resource in Southern California?', MSc thesis, California State University, Northridge (2015).
Asad, Muhammad, and others, 'Green Synthesized *Alnus altissima*-Conjugated $CaCO_3$ Nanoparticles for Diabetes Management', *Frontiers in Chemistry*, 12 (2024).
Ashworth, William B., 'Christianity and the Mechanistic Universe', in *When Science and Christianity Meet*, ed. David Lindberg and Ronald Numbers (Chicago: University of Chicago Press, 2008), pp. 61–84.
Baker, Margaret, *Discovering the Folklore of Plants*, 3rd edn (Boxley: Shire Classics, 2011 [1969]).
Beyond Pesticides, 'Biodiversity in Agriculture and Ecosystem Health' (2024).
——, 'Monoculture in Crop Production Contributes to Biodiversity Loss and Pollinator Decline' (2019).
Bondonno, Nicola P., and others, 'The Cardiovascular Health Benefits of Apples: Whole Fruit vs. Isolated Compounds', *Trends in Food Science & Technology*, 69 (2017), 243–56.
Bongers, Frans, and others, 'Frankincense in Peril', *Nature Sustainability*, 2 (2019), 602–10.

Borisjuk, Nikolai, and others, 'Genetic Modification for Wheat Improvement: From Transgenesis to Genome Editing', *BioMed Research International* (2019), 6216304.

Bucińska, Katarzyna, and others, 'Phytochemical Analysis and Antioxidant Activity of Selected Alder Species', *Plants*, 10 (2021), 2531.

——, 'Polyphenolic Composition and Antioxidant Potential of *Sorbus aucuparia* (Rowan) Berries Compared with Other Fruits', *Open Agriculture Journal*, 17 (2023).

Buffi, Matteo, and others, 'Electrical Signaling in Fungi: Past and Present Challenges', *FEMS Microbiology Reviews*, 49 (2025), fuaf009.

Carson, Rachel, 'Silent Spring', in *Thinking About the Environment: Readings on Politics, Property and the Physical World*, ed. Matthew A. Cahn and Rory O'Brien (London: Routledge, 2015), pp. 150–55.

Charubel, *Grimoire Sympathia: The Workshop of the Infinite*, 2nd edn (London: I-H-O Books, 1906).

Cheal, David, and Jane Leverick, 'Working Magic in Neo-Paganism', *Journal of Ritual Studies*, 13 (1999), 7–19.

Davies, Owen, and Ceri Houlbrook, 'Seeking Protection: Objects of Power', in *Building Magic: Ritual and Re-Enchantment in Post-Medieval Structures*, ed. Owen Davies and Ceri Houlbrook (Cham: Springer International Publishing, 2021), pp. 95–119.

Dietz, S. Theresa, *The Complete Language of Flowers: A Definitive and Illustrated History* (New York: Wellfleet Press, 2020).

Dragicević, Anđela, and others, 'Biological Activity of the Birch Leaf and Bark', *Natural Medicinal Materials*, 42 (2022), 89–105.

Evans, Matthew, *Soil: The Incredible Story of What Keeps the Earth, and Us, Healthy* (London: Murdoch Books, 2022).

Fry, Janis, *The God Tree* (Taunton: Capall Bann Publishing, 2012).

Gaster, Moses, Book Reviews, *Folklore*, 42 (1931), 485–87.

Gorji, Narjes and others, 'Almond, Hazelnut and Walnut, Three Nuts for Neuroprotection in Alzheimer's Disease: A Neuropharmacological Review of Their Bioactive Constituents', *Pharmacological Research*, 129 (2018), 115–27.

Greer, John Michael, *Encyclopedia of Natural Magic* (Woodbury, MN: Llewellyn Publications, 2005).

Harvey, Graham, *Listening People, Speaking Earth: Contemporary Paganism*, 2nd edn (London: C. Hurst and Co., 2007).

Haycock, Marged, 'The Significance of the "Cad Goddau" Tree-List in the Book of Taliesin', in *Celtic Linguistics: Readings in the Brythonic Languages: Festschrift for T. Arwyn Watkins* (Amsterdam: John Benjamins, 1990), pp. 297–331.

Hohman, John George, *The Long Lost Friend*, translated from the German (Harrisburg, PA, 1850).

Holzinger, Felix, and Jean-François Chenot, 'Systematic Review of Clinical Trials Assessing the Effectiveness of Ivy Leaf (*Hedera helix*) for Acute Upper Respiratory Tract Infections', *Evidence-Based Complementary and Alternative Medicine* (2011), 382789.

Hukantaival, Sonja, 'Horse Skulls and "Alder Horse": The Horse as a Depositional Sacrifice in Buildings', *Archaeologia Baltica*, 11 (2009), 350–56.

Hunt, Carter A., and others, 'Setting up Roots: Opportunities for Biocultural Restoration in Recently Inhabited Settings', *Sustainability*, 15 (2023), 2775.

Jaubet, María L., and others, 'Factors Driving the Abundance and Distribution of Microplastics on Sandy Beaches in a Southwest Atlantic Seaside Resort', *Marine Environmental Research*, 171 (2021), 105472.

Kardos, Peter, and others, 'Efficacy and Safety of a Single Ivy Extract Versus Two Herbal Extract Combinations in Patients with Acute Bronchitis: A Multi-Center Randomized, Open-Label Clinical Trial', *Pharmaceuticals* 18(5) (2024), 754.

Lal, R., 'Soil Degradation as a Reason for Inadequate Human Nutrition', *Food Security*, 1 (2009), 45–57.

Leitch, Yuri, *Ogham Grove* (Glastonbury: Yuri Leitch, n.d.).

Leonard, Sophie, and others, 'Microplastics in Human Blood: Polymer Types, Concentrations and Characterisation Using μFTIR', *Environment International*, 188 (2024), 108751.

Lewis, Simon L., and Mark A. Maslin, 'Defining the Anthropocene', *Nature*, 519 (2015), 171–80.

Macy, Joanna, 'The Greening of the Self', in *Spiritual Ecology: The Cry of the Earth*, ed. Llewellyn Vaughan-Lee (Point Reyes, CA: The Golden Sufi Center, 2013), pp. 145–58.

Morgan, Prys, 'A Welsh Snakestone, Its Tradition and Folklore', *Folklore*, 94 (1983), 184–91.

News-Medical, 'Alder Bark May Be Great Source of Anti-Aging and Anti-Disease Natural Antioxidants' (2019).

Oszmiański, Joanna, and others, 'Bioactive Compounds in Sweet Rowanberry Fruits of Interspecific Rowan Crosses', *ResearchGate Preprint* (2023).

Plumwood, Val, 'Gender, Eco-Feminism and the Environment', in *Controversies in Environmental Sociology*, ed. Rob White (New York: Cambridge University Press, 2004), pp. 43–61.

Prayag, Girish, and others, 'Drug or Spirituality Seekers? Consuming Ayahuasca', *Annals of Tourism Research*, 52 (2015), 175–77.

Ragusa, Antonio, and others, 'Plasticenta: First Evidence of Microplastics in Human Placenta', *Environment International*, 146 (2021), 106274.

Richardson, M. J., 'Seed Mycology', *Mycological Research*, 100 (1996), 385–92.

Rop, Kinga, and others, 'Antibacterial Activity of Rowan (*Sorbus aucuparia*) Extracts against Foodborne Pathogens', *Antioxidants*, 10 (2023), 1779.

Ruohonen, Juha, 'A Witch's Coin from Tervo', *Times, Things & Places*, 36 (2011), 344–57.

Šavikin, Elena, and others, 'Evaluating Birch Leaf Infusion as a Functional Herbal Beverage: Anti-Inflammatory and Anti-Adhesive Activities in Urinary Health', *PubMed* (2024).

Shotyk, W., and M. Krachler, 'Contamination of Bottled Waters with Antimony Leaching from Polyethylene Terephthalate (PET) Increases upon Storage', *Environmental Science & Technology*, 41 (2007), 1560–63.

Sibley, Sian, *Unveiling the Green: Working Astrologically, Alchemically and Psychologically with Plants* (Black Lodge Publishing, 2022).

Skinner, Stephen, *Terrestrial Astrology: Divination by Geomancy* (London: Routledge, 1986).

Sonnex, Charmaine, and others, 'Flow, Liminality, and Eudaimonia: Pagan Ritual Practice as a Gateway to a Life with Meaning', *Journal of Humanistic Psychology*, 62 (2022), 233–56.

Spretnak, Charlene, *The Spiritual Dimension of Green Politics* (Santa Fe: Bear Co., 1987).

SpringerLink, 'Fertiliser and Soil Degradation' (2022).

Stapley, Christina, *The Tree Dispensary* (London: Aeon Books, 2021).

Stokstad, R. S., 'Australian Trial of Gene-Edited Wheat Aims for 10% Yield Increase' (Reuters, 2024).

Taylor, Bron, *Dark Green Religion: Nature Spirituality and the Planetary Future* (Berkeley, CA: University of California Press, 2009).

Thwaite, A., 'A History of Amulets in Ten Objects', *Science Museum Group Journal*, 11 (2023).

Tuck, Caroline J., and others, 'Food Intolerances', *Nutrients*, 11 (2019), 1684.

US Geological Survey, 'Nutrients and Eutrophication' (2021).

Valiente, D., *An ABC of Witchcraft Past and Present* (London: Hale, 1989).

Völp, Andreas, and others, 'Ivy Leaves Extract EA 575 in the Treatment of Cough During Acute Respiratory Tract Infections: Meta-Analysis of Double-Blind, Randomized, Placebo-Controlled Trials', *Scientific Reports*, 12 (2022), 20041.

Wen, Ye, and others, 'Medical Empirical Research on Forest Bathing (*Shinrin-Yoku*): A Systematic Review', *Environmental Health and Preventive Medicine*, 24 (2019), 1–21.

Whelchel, Hugh, 'Three Key Bible Passages About Stewardship', *Institute for Faith, Work & Economics* (2023).

White, Robert, *Controversies in Environmental Sociology* (Cambridge: Cambridge University Press, 2004).

www.ingramcontent.com/pod-product-compliance
Lightning Source LLC
Chambersburg PA
CBHW070755230426
43665CB00017B/2371